T0286265

Adaptive Filtering: Concepts, Approaches and Applied Principles

Adaptive Filtering: Concepts, Approaches and Applied Principles

Edited by **Howard Zea**

LANRYE
INTERNATIONAL

New Jersey

Published by Clanrye International,
55 Van Reypen Street,
Jersey City, NJ 07306, USA
www.clanryeinternational.com

Adaptive Filtering: Concepts, Approaches and Applied Principles
Edited by Howard Zea

International Standard Book Number: 978-1-63240-013-0 (Hardback)

Printed in the United States of America.

Contents

Preface

Every book is a source of knowledge and this one is no exception. The idea that led to the conceptualization of this book was the fact that the world is advancing rapidly; which makes it crucial to document the progress in every field. I am aware that a lot of data is already available, yet, there is a lot more to learn. Hence, I accepted the responsibility of editing this book and contributing my knowledge to the community.

This book is a collaboration of researches by renowned experts in the field of adaptive filtering. This system is used to characterize unfamiliar systems in time-variant surroundings. The main purpose of this method is to combine highest convergence speed with maximum accuracy. Every function requires a definite approach which determines the filter composition, the cost function to reduce the judgment error, the adaptive algorithm, and other parameters; and each selection involves certain cost in computational terms, that in any case should consume less time than the time necessary by the application functioning in real-time. This book is a compilation of various aspects related to adaptive filtering and it intends to help researchers in a productive manner.

While editing this book, I had multiple visions for it. Then I finally narrowed down to make every chapter a sole standing text explaining a particular topic, so that they can be used independently. However, the umbrella subject sinews them into a common theme. This makes the book a unique platform of knowledge.

I would like to give the major credit of this book to the experts from every corner of the world, who took the time to share their expertise with us. Also, I owe the completion of this book to the never-ending support of my family, who supported me throughout the project.

Editor

Hirschman Optimal Transform Block LMS Adaptive Filter

Osama Alkhouli, Victor DeBrunner and
Joseph Havlicek

Additional information is available at the end of the chapter

1. Introduction

The HOT is a recently developed discrete unitary transform that uses the orthonormal minimizers of the entropy-based Hirschman uncertainty measure [2]. This measure is different from the energy-based Heisenberg uncertainty measure that is only suited for continuous time signals. The Hirschman uncertainty measure uses entropy to quantify the spread of discrete-time signals in time and frequency [3]. Since the HOT bases are among the minimizers of the uncertainty measure, they have the novel property of being the most compact in discrete time and frequency. The fact that the HOT basis sequences have many zero-valued samples, and their resemblance to the DFT basis sequences, makes the HOT computationally attractive. Furthermore, it has been shown recently that a thresholding algorithm using the HOT yields superior frequency resolution of a pure tone in additive white noise to a similar algorithm based on the DFT [4]. The main theorem in [2] describes a method to generate an $N = K^2$-point orthonormal HOT basis, where K is an integer. A HOT basis sequence of length K^2 is the most compact bases in the time-frequency plane. The 32-point HOT matrix is

$$
\begin{bmatrix}
1\,0\,0 & 1 & 0 & 0 & 1 & 0 & 0 \\
0\,1\,0 & 0 & 1 & 0 & 0 & 1 & 0 \\
0\,0\,1 & 0 & 0 & 1 & 0 & 0 & 1 \\
1\,0\,0\,e^{-j2\pi/3} & 0 & 0 & e^{-j4\pi/3} & 0 & 0 \\
0\,1\,0 & 0 & e^{-j2\pi/3} & 0 & 0 & e^{-j4\pi/3} & 0 \\
0\,0\,1 & 0 & 0 & e^{-j2\pi/3} & 0 & 0 & e^{-j4\pi/3} \\
1\,0\,0\,e^{-j4\pi/3} & 0 & 0 & e^{-j8\pi/3} & 0 & 0 \\
0\,1\,0 & 0 & e^{-j4\pi/3} & 0 & 0 & e^{-j8\pi/3} & 0 \\
0\,0\,1 & 0 & 0 & e^{-j4\pi/3} & 0 & 0 & e^{-j8\pi/3}
\end{bmatrix}
\tag{1}
$$

Equation (1) indicates that the HOT of any sequence is related to the DFT of some polyphase components of the signal. In fact, we called this property the "1 and 1/2 dimensionality" of the HOT in [3]. Consequently, for this chapter, we will use the terms HOT and DFT of the polyphase components interchangeably. The K^2-point HOT requires fewer computations than the K^2-point DFT. We used this computational efficiency of the HOT to implement fast convolution algorithms [5]. When K is a power of 2 integer, then $K^2 \log_2 K$ (complex) multiplications are needed to compute the HOT, which is half that is required when computing the DFT. In this chapter, we use the computational efficiency of the HOT to implement a fast block LMS adaptive filter. The fast block LMS adaptive filter was first proposed [6] to reduce computational complexity. Our proposed HOT block LMS adaptive filter requires less than half of the computations required in the corresponding DFT block LMS adaptive filter. This significant complexity reduction could be important in many real time applications.

The following notations are used throughout this chapter. Nonbold lowercase letters are used for scalar quantities, bold lowercase is used for vectors, and bold uppercase is used for matrices. Nonbold uppercase letters are used for integer quantities such as length or dimensions. The lowercase letter k is reserved for the block index. The lowercase letter n is reserved for the time index. The time and block indexes are put in brackets, whereas subscripts are used to refer to elements of vectors and matrices. The uppercase letter N is reserved for the filter length and the uppercase letter L is reserved for the block length. The superscripts T and H denote vector or matrix transposition and Hermitian transposition, respectively. The subscripts F and H are used to highlight the DFT and HOT domain quantities, respectively. The $N \times N$ identity matrix is denoted by $\mathbf{I}_{N \times N}$ or $\mathbf{I}$. The $N \times N$ zero matrix is denoted by $\mathbf{0}_{N \times N}$. The linear and circular convolutions are denoted by $*$ and $\star$, respectively. Diag $[\mathbf{v}]$ denotes the diagonal matrix whose diagonal elements are the elements of the vector $\mathbf{v}$.

2. The relation between the HOT and DFT in a matrix from

The algorithm that we proposing is best analyzed if the relation between the HOT and DFT is presented in matrix form. This matrix form is shown in Figure 1, where $\mathbf{I}_0, \mathbf{I}_1,..., \mathbf{I}_{K-1}$ are $K \times K^2$ matrices such that multiplication of a vector with $\mathbf{I}_i$ produces the ith polyphase component of the vector. The matrix $\mathbf{I}_K$ is formed from $\mathbf{I}_0, \mathbf{I}_1,..., \mathbf{I}_{K-1}$, i.e.,

$$
\mathbf{I}_K = \begin{bmatrix} \mathbf{I}_0 \\ \mathbf{I}_1 \\ \vdots \\ \mathbf{I}_{K-2} \\ \mathbf{I}_{K-1} \end{bmatrix}. \tag{2}
$$

Since the rows of $\left\{ \mathbf{I}_i \right\}$ are taken from the rows of the $K^2 \times K^2$ identity matrix, multiplications with such matrices does not impose any computational burden. For the special case $K = 3$, we have

$$
\mathbf{I}_0 = \begin{bmatrix} 1\,0\,0\,0\,0\,0\,0\,0\,0 \\ 0\,0\,0\,1\,0\,0\,0\,0\,0 \\ 0\,0\,0\,0\,0\,0\,1\,0\,0 \end{bmatrix}, \tag{3}
$$

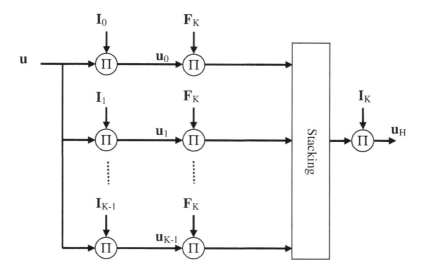

Figure 1. The Relation between HOT and DFTs of the polyphase components.

$$\mathbf{I}_1 = \begin{bmatrix} 0\,1\,0\,0\,0\,0\,0\,0\,0 \\ 0\,0\,0\,0\,1\,0\,0\,0\,0 \\ 0\,0\,0\,0\,0\,0\,0\,1\,0 \end{bmatrix}, \tag{4}$$

$$\mathbf{I}_2 = \begin{bmatrix} 0\,0\,1\,0\,0\,0\,0\,0\,0 \\ 0\,0\,0\,0\,0\,1\,0\,0\,0 \\ 0\,0\,0\,0\,0\,0\,0\,0\,1 \end{bmatrix}. \tag{5}$$

The K^2-point HOT matrix is denoted by $\mathbf{H}$. It satisfies the following:

$$\mathbf{H}\mathbf{H}^H = K\mathbf{I}_{K^2 \times K^2}, \tag{6}$$

$$\mathbf{H} = \mathbf{H}^T. \tag{7}$$

3. Convolution using the HOT

In this section, the "HOT convolution," a relation between the HOT of two signals and their circular convolution, is derived. Let u and w be two signals of length K^2. The circular convolution of the signals is $y = w \star u$. In the DFT domain, the convolution is given by the pointwise multiplication of the respective DFTs of the signals, i.e., $y_F(k) = w_F(k)u_F(k)$. A similar relation in the HOT domain can be readily found through the relation between the DFT and HOT. The DFT of u can be written as

$$u_F(k) = \sum_{n=0}^{K^2-1} u(n)\, e^{-j\frac{2\pi}{K^2}kn}$$

$$= \sum_{i=0}^{K-1} e^{-j\frac{2\pi}{K^2}ki} \sum_{l=0}^{K-1} u(lK+i)\, e^{-j\frac{2\pi}{K}kl}. \tag{8}$$

The signal $u(lK+i)$, denoted by $u_i(l)$, is the ith polyphase component of $u(n)$ with DFT given by

$$u_{iF}(k) = \sum_{l=0}^{K-1} u_i(l)\, e^{-j\frac{2\pi}{K}kl}. \tag{9}$$

Therefore, the DFT of the signal u can be written in terms of the DFTs of the polyphase components, or the HOT of u. The relation between the HOT and the DFTs of the polyphase components is descried in Figure 1. Equation (8) may be written as

$$u_F(k) = \sum_{i=0}^{K-1} e^{-j\frac{2\pi}{K^2}ki} u_{iF}(k). \tag{10}$$

Define the diagonal matrix

$$\mathbf{D}_{i,j}(k) = \begin{bmatrix} e^{-j\frac{2\pi}{K^2}ki} & 0 & \cdots & 0 \\ 0 & e^{-j\frac{2\pi}{K^2}k(i+1)} & \cdots & 0 \\ \vdots & \vdots & \ddots & \vdots \\ 0 & 0 & \cdots & e^{-j\frac{2\pi}{K^2}kj} \end{bmatrix} \tag{11}$$

Then the DFT of the signal can be written in a matrix form

$$\mathbf{u}_F = \sum_{i=0}^{K-1} \mathbf{D}_{0,K^2-1}(i) \begin{bmatrix} \mathbf{F}_K \\ \mathbf{F}_K \\ \vdots \\ \mathbf{F}_K \end{bmatrix} \mathbf{u}_i. \tag{12}$$

The above is the desired relation between the DFT and HOT. It should be noted that equation (12) represents a radix-K FFT algorithm which is less efficient than the radix-2 FFT algorithm. Therefore, HOT convolution is expected to be less efficient than DFT convolution. Now, we can use equation (12) to transform $\mathbf{y}_F = \mathbf{w}_F \otimes \mathbf{u}_F$ into the HOT domain. The symbol $\otimes$ indicates pointwise matrix multiplication and, throughout this discussion, pointwise matrix multiplication takes a higher precedence than conventional matrix multiplication. We have that

$$\sum_{i=0}^{K-1} \mathbf{D}_{0,K^2-1}(i) \begin{bmatrix} \mathbf{F}_K \\ \mathbf{F}_K \\ \vdots \\ \mathbf{F}_K \end{bmatrix} \mathbf{y}_i = \sum_{i=0}^{K-1}\sum_{j=0}^{K-1} \mathbf{D}_{0,K^2-1}(i+j) \begin{bmatrix} \mathbf{F}_K \mathbf{w}_i \\ \mathbf{F}_K \mathbf{w}_i \\ \vdots \\ \mathbf{F}_K \mathbf{w}_i \end{bmatrix} \otimes \begin{bmatrix} \mathbf{F}_K \mathbf{u}_j \\ \mathbf{F}_K \mathbf{u}_j \\ \vdots \\ \mathbf{F}_K \mathbf{u}_j \end{bmatrix}. \tag{13}$$

The above matrix equation can be separated into a system of K equations

$$\sum_{i=0}^{K-1} \mathbf{D}_{rK,(r+1)K-1}(i)\mathbf{F}_K\mathbf{y}_i = \sum_{i=0}^{K-1}\sum_{j=0}^{K-1} \mathbf{D}_{rK,(r+1)K-1}(i+j)\,(\mathbf{F}_K\mathbf{w}_i)\otimes\left(\mathbf{F}_K\mathbf{w}_j\right), \tag{14}$$

where $r = 0, 1, \ldots, K-1$. Since

$$\mathbf{D}_{rK,(r+1)K-1}(i) = e^{-j\frac{2\pi}{K}ri}\mathbf{D}_{0,K-1}(i), \tag{15}$$

the HOT of the output can be obtained by solving the following set of K matrix equations:

$$\sum_{i=0}^{K-1} e^{-j\frac{2\pi}{K}ri}\mathbf{D}_{0,K-1}(i)\mathbf{F}_K\mathbf{y}_i = \sum_{i=0}^{K-1}\sum_{j=0}^{K-1} e^{-j\frac{2\pi}{K}r(i+j)}\mathbf{D}_{0,K-1}(i+j)\,(\mathbf{F}_K\mathbf{w}_i)\otimes\left(\mathbf{F}_K\mathbf{u}_j\right). \tag{16}$$

Since the DFT matrix is unitary, the solution of equation (16) can be expressed as

$$\mathbf{D}_{0,K-1}(s)\mathbf{F}_K\mathbf{y}_s = \frac{1}{K}\sum_{r=0}^{K-1}\sum_{i=0}^{K-1}\sum_{j=0}^{K-1} e^{j\frac{2\pi}{K}r(s-(i+j))}\mathbf{D}_{0,K-1}(i+j)\,(\mathbf{F}_K\mathbf{w}_i)\otimes\left(\mathbf{F}_K\mathbf{u}_j\right), \tag{17}$$

where

$$\mathbf{F}_K\mathbf{y}_s = \frac{1}{K}\sum_{r=0}^{K-1}\sum_{i=0}^{K-1}\sum_{j=0}^{K-1} e^{j\frac{2\pi}{K}r(i+j-s)}\mathbf{D}_{0,K-1}(i+j-s)\,(\mathbf{F}_K\mathbf{w}_i)\otimes\left(\mathbf{F}_K\mathbf{u}_j\right). \tag{18}$$

Moreover, as

$$\sum_{r=0}^{K-1} e^{j\frac{2\pi}{K}r(i+j-s)} = K\delta(i+j-s), \tag{19}$$

where $\delta(n)$ denotes the periodic Kronecker delta of periodicity K, equation (18) can be simplified to

$$\mathbf{F}_K\mathbf{y}_s = \sum_{i=0}^{K-1}\sum_{j=0}^{K-1} \delta(i+j-s)\mathbf{D}_{0,K-1}(i+j-s)\,(\mathbf{F}_K\mathbf{w}_i)\otimes\left(\mathbf{F}_K\mathbf{u}_j\right), \tag{20}$$

where $s = 0, 1, 2, \ldots, K-1$. The pointwise matrix multiplication in equation equation (20) can be converted into conventional matrix multiplication if we define $\mathbf{W}_i$ as the diagonal matrix for $\mathbf{F}_K\mathbf{w}_i$. We have then that

$$\mathbf{F}_K \mathbf{y}_s = \sum_{i=0}^{K-1} \sum_{j=0}^{K-1} \delta(i+j-s) \mathbf{D}_{0,K-1}(i+j-s) \mathbf{W}_i \mathbf{F}_K \mathbf{u}_j. \tag{21}$$

Combining the above K equations into one matrix equation, the HOT convolution can be written as

$$
\begin{bmatrix}
\mathbf{F}_K \mathbf{y}_0 \\
\mathbf{F}_K \mathbf{y}_1 \\
\mathbf{F}_K \mathbf{y}_2 \\
\vdots \\
\mathbf{F}_K \mathbf{y}_{K-2} \\
\mathbf{F}_K \mathbf{y}_{K-1}
\end{bmatrix}
=
\begin{bmatrix}
\mathbf{W}_0 & \mathbf{DW}_{K-1} & \mathbf{DW}_{K-2} & \cdots & \mathbf{DW}_2 & \mathbf{DW}_1 \\
\mathbf{W}_1 & \mathbf{W}_0 & \mathbf{W}_{K-1} & \cdots & \mathbf{DW}_3 & \mathbf{DW}_2 \\
\mathbf{W}_2 & \mathbf{W}_1 & \mathbf{W}_0 & \cdots & \mathbf{DW}_4 & \mathbf{DW}_3 \\
\vdots & \vdots & \vdots & \ddots & \vdots & \vdots \\
\mathbf{W}_{K-2} & \mathbf{W}_{K-3} & \mathbf{W}_{K-4} & \cdots & \mathbf{W}_0 & \mathbf{DW}_{K-1} \\
\mathbf{W}_{K-1} & \mathbf{W}_{K-2} & \mathbf{W}_{K-3} & \cdots & \mathbf{W}_1 & \mathbf{W}_0
\end{bmatrix}
\begin{bmatrix}
\mathbf{F}_K \mathbf{u}_0 \\
\mathbf{F}_K \mathbf{u}_1 \\
\mathbf{F}_K \mathbf{u}_2 \\
\vdots \\
\mathbf{F}_K \mathbf{u}_{K-2} \\
\mathbf{F}_K \mathbf{u}_{K-1}
\end{bmatrix}
\tag{22}
$$

where

$$
\mathbf{D} =
\begin{bmatrix}
1 & 0 & \cdots & & 0 \\
0 & e^{-j\frac{2\pi}{K^2}} & \cdots & & 0 \\
\vdots & \vdots & \ddots & & \vdots \\
0 & 0 & \cdots & & e^{-j\frac{2\pi}{K^2}(K-1)}
\end{bmatrix}
\tag{23}
$$

Notice that the square matrix in equation (22) is arranged in a block Toeplitz structure.

A better understanding of this result may be obtained by comparing equation (22) with the K-point circular convolution

$$
\begin{bmatrix}
y_0 \\
y_1 \\
y_2 \\
\vdots \\
y_{K-2} \\
y_{K-1}
\end{bmatrix}
=
\begin{bmatrix}
w_0 & w_{K-1} & w_{K-2} & \cdots & w_2 & w_1 \\
w_1 & w_0 & w_{K-1} & \cdots & w_3 & w_2 \\
w_2 & w_1 & w_0 & \cdots & w_4 & w_3 \\
\vdots & \vdots & \vdots & \ddots & \vdots & \vdots \\
w_{K-2} & w_{K-3} & w_{K-4} & \cdots & w_0 & w_{K-1} \\
w_{K-1} & w_{K-2} & w_{K-3} & \cdots & w_1 & w_0
\end{bmatrix}
\begin{bmatrix}
u_0 \\
u_1 \\
u_2 \\
\vdots \\
u_{K-2} \\
u_{K-1}
\end{bmatrix}.
\tag{24}
$$

The square matrix in equation (24) is also Toeplitz. However, equation (24) is a pure time domain result, whereas equation (22) is a pure HOT domain relation, which may be interpreted in terms of both the time domain and the DFT domain features. This fact can be explained in terms of fact that the HOT basis is optimal in the sense of the entropic joint time-frequency uncertainty measure $H_p(u) = pH(u) + (1-p)H(u_F)$ for all $0 \le p \le 1$. Before moving on to the computational complexity analysis of HOT convolution, we make the same observations about the term $\mathbf{DF}_K \mathbf{w}_i$ appearing in equation (22). This term is the complex conjugate of the DFT of the upside down flipped i^{th} polyphase component of w.

It should be noted that equation (22) does not show explicitly the HOT of $u(n)$ and $w(n)$. However, the DFT of the polyphase components that are shown explicitly in equation (22) are related to the HOT of the corresponding signal as shown in Figure. 1. For example, the 0th polyphase component of the output is given by

$$\mathbf{y}_0(k) = \mathbf{F}_K^{-1}\mathbf{I}_0\mathbf{w}_H(k) \otimes \mathbf{I}_0\mathbf{u}_H(k) + \mathbf{F}_K^{-1}\mathbf{D}\sum_{i=1}^{K-1}\mathbf{I}_{K-i}\mathbf{w}_H(k) \otimes \mathbf{I}_i\mathbf{u}_H(k). \tag{25}$$

Next, we examine the computational complexity of HOT convolution. To find the HOT of the two signals w and u, $2K^2\log_2 K$ multiplications are required. Multiplication with the diagonal matrix D requires $K(K-1)$ multiplications. Finally, the matrix multiplication requires K^3 scalar multiplications. Therefore, the total number of multiplications required is $2K^2\log_2 K + K^3 + K^2 - K$. Thus, computation of the output y using the HOT requires $K^3 + 3K^2\log_2 K + K^3 + K^2 - K$ multiplications, which is more than $6K^2\log_2 K + K^2$ as required by the DFT. When it is required to calculate only one polyphase component of the output, only $K^2 + 2K^2\log_2 K + K\log_2 K$ multiplications are necessary. Asymptotically in K, we see that the HOT could be three times more efficient than the DFT.

4. Development of the basic algorithm

In the block adaptive filter, the adaptation proceeds block-by-block with the weight update equation

$$\mathbf{w}(k+1) = \mathbf{w}(k) + \frac{\mu}{L}\sum_{i=0}^{L-1}\mathbf{u}(kL+i)e(kL+i), \tag{26}$$

where $d(n)$ and $y(n)$ are the desired and output signals, respectively, $\mathbf{u}(n)$ is the tap-input vector, L is the block length or the filter length, and $e(n) = d(n) - y(n)$ is the filter error. The DFT is commonly used to efficiently calculate the output of the filter and the sum in the update equation. Since the HOT is more efficient than the DFT when it is only required to calculate one polyphase component of the output, the block LMS algorithm equation (26) is modified such that only one polyphase component of the error in the kth block is used to update the filter weights. For reasons that will become clear later, the filter length L is chosen such that $L = K^2/2$. With this modification, equation (26) becomes

$$\mathbf{w}(k+1) = \mathbf{w}(k) + \frac{2\mu}{K}\sum_{i=0}^{K/2-1}\mathbf{u}(kL+iK+j)e(kL+iK+j). \tag{27}$$

Since the DFT is most efficient when the length of the filter is equal to the block length [7], this will be assumed in equation (27). The parameter j determines which polyphase component of the error signal is being used in the adaptation. This parameter can be changed from block to block. If $j = 0$, the output can be computed using the HOT as in equation (25). A second convolution is needed to compute the sum in equation (27). This sum contains only one polyphase component of the error. If this vector is up-sampled by K, the sum is just a convolution between the input vector and the up-sampled error vector. Although all the polyphase components are needed in the sum, the convolution can be computed by the HOT with the same computational complexity as the first convolution since only one polyphase component of the error vector is non-zero.

The block adaptive filter that implements the above algorithm is called the HOT block LMS adaptive filter and is shown in Figure 2. The complete steps of this new, efficient, adaptive algorithm are summarized below:

(a) Append the weight vector with $K^2/2$ zeros (the resulting vector is now K^2 points long as required in the HOT definition) and find its HOT.

(b) Compute the HOT of the input vector

$$\left[u\left((k-1)\tfrac{K^2}{2}\right) \cdots u\left(k\tfrac{K^2}{2}\right) u\left(k\tfrac{K^2}{2}+1\right) \cdots u\left((k+1)\tfrac{K^2}{2}-1\right) \right]^T. \tag{28}$$

Note that this vector contains the input samples for the current and previous blocks.

(c) Use the inverse HOT and equation (22) to calculate the jth polyphase component of the circular convolution. The jth polyphase component of the output can be found by discarding the first half of the jth polyphase component of the circular convolution.

(d) Calculate the jth polyphase component of the error, insert a block of $K/2$ zeros, up-sample by K, then calculate its HOT.

(e) Circularly flip the vector in (b) and then compute its HOT.

(f) Compute the sum in the update equation using equation (22). This sum is the first half of the elements of the circular convolution between the vectors in parts (e) and (d).

5. Computational complexity analysis

In this section, we analyze the computational cost of the algorithm and compare it to that of the DFT block adaptive algorithm. Parts (a), (b), and (e) require $3K^2 \log_2 K$ multiplications. Part (c) requires $K \log_2 K + K^2$. Part (d) requires $K \log_2 K$ multiplications, and part (f) requires $K^2 + K^2 \log_2 K$ multiplications. The total number of multiplications is thus $4K^2 \log_2 K + 2K \log_2 K + 2K^2$. The corresponding DFT block adaptive algorithm requires $10K^2 \log_2 K + 2K^2$ multiplications — asymptotically more than twice as many. Therefore, by using only one polyphase component for the adaptation in a block, the computational cost can be reduced by a factor of 2.5. While this complexity reduction comes at the cost of not using all available information, the proposed algorithm provides better estimates than the LMS filter. The reduction of the computational complexity in this algorithm comes from using the polyphase components of the input signal to calculate one polyphase component of the output via the HOT.

It is worth mentioning that the fast exact LMS (FELMS) adaptive algorithm [8] also reduces the computational complexity by finding the output by processing the polyphase components of the input. However, the computational complexity reduction of the FELMS algorithm is less than that found in the DFT and HOT block adaptive algorithms because the FELMS algorithm is designed to have exact mathematical equivalence to, and hence the same convergence properties as, the conventional LMS algorithm. Comparing the HOT block LMS algorithm with the block LMS algorithms described in Chapter 3, the HOT filter performs computationally better.

The multiplication counts for both the DFT block and HOT block LMS algorithms are plotted in Figure 3. The HOT block LMS adaptive filter is always more efficient than the DFT block

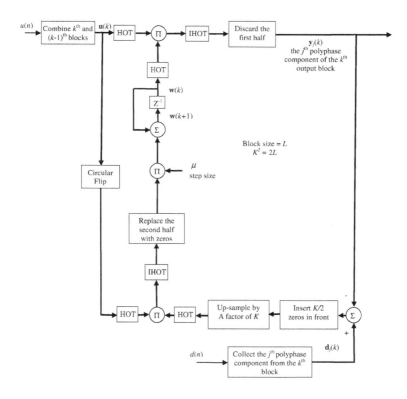

Figure 2. HOT block LMS adaptive filter.

LMS adaptive filter and the asymptotic ratio between their computational cost is almost reached at small filter lengths. The computational complexity of the HOT filter can be further improved by relating the HOT of the circularly flipped vector in step (e) to the HOT of the vector in step (b). Another possibility to reduce the computational cost of the HOT block algorithm is by removing the gradient constraint in the filter weight update equation as has been done in the unconstrained DFT block LMS algorithm [9].

6. Convergence analysis in the time domain

In this section, we analyze the convergence of the HOT block LMS algorithm in the time domain. We assume throughout that the step size is small. The HOT block LMS filter minimizes the cost

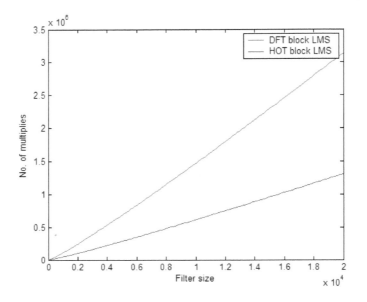

Figure 3. Multiplication counts for both the DFT block and HOT block LMS algorithms.

$$\hat{\zeta} = \frac{2}{K} \sum_{i=0}^{\frac{K}{2}-1} \left| e(kL + iK + j) \right|^2,$$ (29)

which is the average of the squared errors in the jth polyphase error component. From statistical LMS theory [10], the block LMS algorithm can be analyzed using the stochastic difference equation [10]

$$\epsilon_T(k+1) = \left(\mathbf{I} - \mu\mathbf{\Lambda} \right) \epsilon_T(k) + \phi(k),$$ (30)

where

$$\phi(k) = -\frac{\mu}{L} \mathbf{V}^H \sum_{i=0}^{L-1} \mathbf{u}(kL + i) \, e^o(kL + i)$$ (31)

is the driving force of for the block LMS algorithm [10]. we found that the HOT block LMS algorithm has the following driving force

$$\phi_{\text{HOT}}(k) = -\frac{2\mu}{K} \mathbf{v}^H \sum_{i=0}^{\frac{K}{2}-1} \mathbf{u}(kL + iK + j) \, e^o(kL + iK + j). \tag{32}$$

It is easily shown that

$$E\phi_{\text{HOT}}(k) = 0, \tag{33}$$

$$E\phi_{\text{HOT}}(k)\phi_{\text{HOT}}^H(k) = \frac{2\mu^2 J_{\min}\boldsymbol{\Lambda}}{K}. \tag{34}$$

The mean square of the lth component of equation (34) is given by

$$E\,|\epsilon_l(k)|^2 = \frac{2\mu \frac{J_{\min}}{K}}{2 - \mu\lambda_l} + (1 - \mu\lambda_l)^{2k} \left(|\epsilon_l(0)|^2 - \frac{2\mu \frac{J_{\min}}{K}}{2 - \mu\lambda_l} \right), \tag{35}$$

where λ_l is the lth eigenvalue of the input autocorrelation matrix. Therefore, the average time constant of the HOT block LMS algorithm is given by

$$\tau = \frac{L^2}{2\mu \sum_{l=1}^{L} \lambda_l}. \tag{36}$$

The misadjustment can be calculated directly and is given by

$$M = \frac{\sum_{l=1}^{L} \lambda_l E\,|\epsilon_l(\infty)|^2}{J_{\min}}. \tag{37}$$

Using equation (30), one may find $E|\epsilon_l(\infty)|^2$ and substitute the result into equation (37). The misadjustment of the HOT block LMS filter is then given by

$$M = \frac{\mu}{K} \sum_{l=1}^{L} \lambda_l. \tag{38}$$

Thus, the average time constant of the HOT block LMS filter is the same as that of the DFT block LMS filter [1]. However, the HOT block LMS filter has K times higher misadjustment than the DFT block LMS algorithm [2].

The HOT and DFT block LMS algorithms were simulated using white noise inputs. The desired signal was generated using the linear model $d(n) = w^o(n) * u(n) + e^o(n)$, where $e^o(n)$ is the measurement white gaussian noise with variance 10^{-4} and $W^o(z) = 1 + 0.5z^{-1} -$

---

[1] The average time constant of the DFT block LMS filter is [10] $\tau = L^2/2\mu \sum_{l=1}^{L} \lambda_l$.
[2] The misadjustment of the DFT block LMS algorithm is [10] $M = \frac{\mu}{K^2} \sum_{l=1}^{L} \lambda_l$.

$0.25z^{-2} + 0.03z^{-3} + 0.1z^{-4} + 0.002z^{-5} - 0.01z^{-6} + 0.007z^{-7}$. The learning curves are shown in Figure 4 with the learning curve of the conventional LMS algorithm. The step sizes of all algorithms were chosen to be the same. The higher mean square error of the HOT algorithm, compared to the DFT algorithm, shows the trade-off for complexity reduction by more than half. As expected the HOT and DFT block LMS algorithms converge at the same rate.

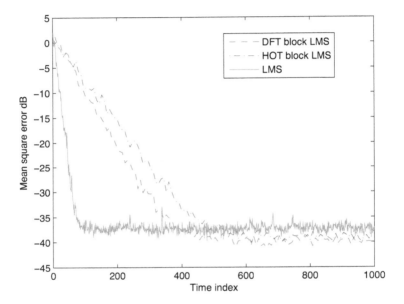

Figure 4. Learning curves of the DFT and HOT block LMS algorithms with the conventional LMS filter.

7. Convergence analysis in the HOT domain

Let $u(n)$ be the input to the adaptive filter and

$$\hat{\mathbf{w}}(k) = \left[w_0(k) \; w_1(k) \; \cdots \; w_{\frac{K^2}{2}-1}(k) \right]^T \tag{39}$$

be the tap-weight vector of the adaptive filter, where k is the block index. Define the extended tap-weight vector

$$\mathbf{w}(k) = \left[\hat{\mathbf{w}}^T(k) \; 0 \; 0 \; \cdots \; 0 \right]^T \tag{40}$$

and the tap-input vector

$$\mathbf{u}(k) = \left[u\left((k-1)\tfrac{K^2}{2}\right) \cdots u\left(k\tfrac{K^2}{2}\right) u\left(k\tfrac{K^2}{2}+1\right) \cdots u\left((k+1)\tfrac{K^2}{2}-1\right) \right]^T. \qquad (41)$$

Denote the HOT transforms of $\mathbf{u}(k)$ and $\mathbf{w}(k)$ by $\mathbf{u}_H(k) = \mathbf{H}\mathbf{u}(k)$ and $\mathbf{w}_H(k) = \mathbf{H}\mathbf{w}(k)$, respectively, where $\mathbf{H}$ is the HOT matrix. The 0th polyphase component of the circular convolution of $\mathbf{u}(k)$ and $\mathbf{w}(k)$ is given by

$$\mathbf{F}_K\mathbf{y}_0(k) = \mathbf{F}_K\mathbf{w}_0(k) \otimes \mathbf{F}_K\mathbf{u}_0(k) + \mathbf{D}\sum_{i=1}^{K-1}\mathbf{F}_K\mathbf{w}_{K-i}(k) \otimes \mathbf{F}_K\mathbf{u}_i(k). \qquad (42)$$

Using $\mathbf{F}_K\mathbf{u}_i(k) = \mathbf{I}_i\mathbf{H}\mathbf{u}(k) = \mathbf{I}_i\mathbf{u}_H(k)$, equation (42) can be written in terms of the HOT of u(k) and w(k). The result is given by

$$\mathbf{F}_K\mathbf{y}_0(k) = \mathbf{I}_0\mathbf{w}_H(k) \otimes \mathbf{I}_0\mathbf{u}_H(k) + \mathbf{D}\sum_{i=1}^{K-1}\mathbf{I}_{K-i}\mathbf{w}_H(k) \otimes \mathbf{I}_i\mathbf{u}_H(k). \qquad (43)$$

The 0th polyphase component of the linear convolution of $\hat{\mathbf{w}}(k)$ and $\mathbf{u}(n)$, the output of the adaptive filter in the kth block, is given by the last $K/2$ elements of $\mathbf{y}_0(k)$. Let the desired signal be $d(n)$ and define the extended 0th polyphase component of the desired signal in the kth block as

$$\mathbf{d}_0(k) = \begin{bmatrix} \mathbf{0}_{\frac{K}{2}} \\ \hat{\mathbf{d}}_0(k) \end{bmatrix}. \qquad (44)$$

The extended 0th polyphase component of error signal in the kth block is given by

$$\mathbf{e}_0(k) = \begin{bmatrix} \mathbf{0}_{\frac{K}{2}} \\ \hat{\mathbf{e}}_0(k) \end{bmatrix} = \begin{bmatrix} \mathbf{0}_{\frac{K}{2}} \\ \hat{\mathbf{d}}_0(k) \end{bmatrix} - \begin{bmatrix} \mathbf{0}_{\frac{K}{2}\times\frac{K}{2}} & \mathbf{0}_{\frac{K}{2}\times\frac{K}{2}} \\ \mathbf{0}_{\frac{K}{2}\times\frac{K}{2}} & \mathbf{I}_{\frac{K}{2}\times\frac{K}{2}} \end{bmatrix} \mathbf{F}_K^{-1}$$

$$\times \left[\mathbf{I}_0\mathbf{w}_H(k) \otimes \mathbf{I}_0\mathbf{u}_H(k) + \mathbf{D}\sum_{i=1}^{K-1}\mathbf{I}_{K-i}\mathbf{w}_H(k) \otimes \mathbf{I}_i\mathbf{u}_H(k) \right]. \qquad (45)$$

Multiplying equation (45) by the DFT matrix yields

$$\mathbf{F}_K\mathbf{e}_0(k) = \mathbf{F}_K\begin{bmatrix} \mathbf{0}_{\frac{K}{2}} \\ \hat{\mathbf{d}}_0(k) \end{bmatrix} - \mathbf{F}_K\begin{bmatrix} \mathbf{0}_{\frac{K}{2}\times\frac{K}{2}} & \mathbf{0}_{\frac{K}{2}\times\frac{K}{2}} \\ \mathbf{0}_{\frac{K}{2}\times\frac{K}{2}} & \mathbf{I}_{\frac{K}{2}\times\frac{K}{2}} \end{bmatrix} \mathbf{F}_K^{-1}$$

$$\times \left[\mathbf{I}_0\mathbf{w}_H(k) \otimes \mathbf{I}_0\mathbf{u}_H(k) + \mathbf{D}\sum_{i=1}^{K-1}\mathbf{I}_{K-i}\mathbf{w}_H(k) \otimes \mathbf{I}_i\mathbf{u}_H(k) \right]. \qquad (46)$$

Define $\mathbf{u}_H^c(k) = \mathbf{H}\mathbf{u}^c(k)$, where $\mathbf{u}^c(k)$ is the circularly shifted version of $\mathbf{u}(k)$. The adaptive filter update equation in the kth block is given by

$$\mathbf{w}_H(k+1) = \mathbf{w}_H(k) + \mu\,\mathbf{H}\begin{bmatrix} \mathbf{I}_{\frac{K^2}{2}\times\frac{K^2}{2}} & \mathbf{0}_{\frac{K^2}{2}\times\frac{K^2}{2}} \\ \mathbf{0}_{\frac{K^2}{2}\times\frac{K^2}{2}} & \mathbf{0}_{\frac{K^2}{2}\times\frac{K^2}{2}} \end{bmatrix}\mathbf{H}^{-1}\boldsymbol{\phi}_H(k), \tag{47}$$

where $\boldsymbol{\phi}_H(k)$ is found from

$$\begin{bmatrix} \mathbf{I}_0\boldsymbol{\phi}_H(k) \\ \mathbf{I}_1\boldsymbol{\phi}_H(k) \\ \mathbf{I}_2\boldsymbol{\phi}_H(k) \\ \vdots \\ \mathbf{I}_{K-2}\boldsymbol{\phi}_H(k) \\ \mathbf{I}_{K-1}\boldsymbol{\phi}_H(k) \end{bmatrix} = \begin{bmatrix} \mathbf{F}_K\mathbf{e}_0(k) \\ \mathbf{F}_K\mathbf{e}_0(k) \\ \mathbf{F}_K\mathbf{e}_0(k) \\ \vdots \\ \mathbf{F}_K\mathbf{e}_0(k) \\ \mathbf{F}_K\mathbf{e}_0(k) \end{bmatrix} \otimes \begin{bmatrix} \mathbf{I}_0\mathbf{u}_H^c(k) \\ \mathbf{I}_1\mathbf{u}_H^c(k) \\ \mathbf{I}_2\mathbf{u}_H^c(k) \\ \vdots \\ \mathbf{I}_{K-2}\mathbf{u}_H^c(k) \\ \mathbf{I}_{K-1}\mathbf{u}_H^c(k) \end{bmatrix}, \tag{48}$$

as

$$\boldsymbol{\phi}_H(k) = \mathbf{I}_K^{-1}\begin{bmatrix} \mathbf{F}_K\mathbf{e}_0(k) \\ \mathbf{F}_K\mathbf{e}_0(k) \\ \mathbf{F}_K\mathbf{e}_0(k) \\ \vdots \\ \mathbf{F}_K\mathbf{e}_0(k) \\ \mathbf{F}_K\mathbf{e}_0(k) \end{bmatrix} \otimes \begin{bmatrix} \mathbf{I}_0\mathbf{u}_H^c(k) \\ \mathbf{I}_1\mathbf{u}_H^c(k) \\ \mathbf{I}_2\mathbf{u}_H^c(k) \\ \vdots \\ \mathbf{I}_{K-2}\mathbf{u}_H^c(k) \\ \mathbf{I}_{K-1}\mathbf{u}_H^c(k) \end{bmatrix}. \tag{49}$$

Finally, the HOT block LMS filter in the HOT domain can be written as

$$\mathbf{w}_H(k+1) = \mathbf{w}_H(k)$$

$$+ \mu\,\mathbf{H}\begin{bmatrix} \mathbf{I}_{\frac{K^2}{2}\times\frac{K^2}{2}} & \mathbf{0}_{\frac{K^2}{2}\times\frac{K^2}{2}} \\ \mathbf{0}_{\frac{K^2}{2}\times\frac{K^2}{2}} & \mathbf{0}_{\frac{K^2}{2}\times\frac{K^2}{2}} \end{bmatrix}\mathbf{H}^{-1}\mathbf{I}_K^{-1}\begin{bmatrix} \mathbf{F}_K\mathbf{e}_0(k) \\ \mathbf{F}_K\mathbf{e}_0(k) \\ \mathbf{F}_K\mathbf{e}_0(k) \\ \vdots \\ \mathbf{I}_{K-1}\mathbf{e}_0(k) \\ \mathbf{I}_{K-1}\mathbf{e}_0(k) \end{bmatrix} \otimes \begin{bmatrix} \mathbf{I}_0\mathbf{u}_H^c(k) \\ \mathbf{I}_1\mathbf{u}_H^c(k) \\ \mathbf{I}_2\mathbf{u}_H^c(k) \\ \vdots \\ \mathbf{I}_{K-2}\mathbf{u}_H^c(k) \\ \mathbf{I}_{K-1}\mathbf{u}_H^c(k) \end{bmatrix}. \tag{50}$$

Next, we investigate the convergence properties of equation (50). we assume the following linear statistical model for the desired signal:

$$d(n) = w^o(n) * u(n) + e^o(n), \tag{51}$$

where w^o is the impulse response of the Wiener optimal filter and $e^o(n)$ is the irreducible estimation error, which is white noise and statistically independent of the adaptive filter input. The above equation can be written in the HOT domain form

$$\begin{bmatrix} \mathbf{0}_{\frac{K}{2}} \\ \hat{\mathbf{d}}_0(k) \end{bmatrix} = \begin{bmatrix} \mathbf{0}_{\frac{K}{2} \times \frac{K}{2}} & \mathbf{0}_{\frac{K}{2} \times \frac{K}{2}} \\ \mathbf{0}_{\frac{K}{2} \times \frac{K}{2}} & \mathbf{I}_{\frac{K}{2} \times \frac{K}{2}} \end{bmatrix} \mathbf{F}_K^{-1}$$

$$\times \left[\mathbf{I}_0 \mathbf{w}_H^o(k) \otimes \mathbf{I}_0 \mathbf{u}_H(k) + \mathbf{D} \sum_{i=1}^{K-1} \mathbf{I}_{K-i} \mathbf{w}_H^o(k) \otimes \mathbf{I}_i \mathbf{u}_H(k) + \mathbf{F}_K \mathbf{e}_0^o(k) \right]. \qquad (52)$$

This form will be useful to obtain the stochastic difference equation that describes the convergence of the adaptive algorithm. Using the above equation to replace the desired signal in equation (46), we have

$$\mathbf{F}_K \mathbf{e}_0(k) = \mathbf{F}_K \begin{bmatrix} \mathbf{0}_{\frac{K}{2} \times \frac{K}{2}} & \mathbf{0}_{\frac{K}{2} \times \frac{K}{2}} \\ \mathbf{0}_{\frac{K}{2} \times \frac{K}{2}} & \mathbf{I}_{\frac{K}{2} \times \frac{K}{2}} \end{bmatrix} \mathbf{F}_K^{-1}$$

$$\times \left[\mathbf{I}_0 \boldsymbol{\epsilon}_H(k) \otimes \mathbf{I}_0 \mathbf{u}_H(k) + \mathbf{D} \sum_{i=1}^{K-1} \mathbf{I}_{K-i} \boldsymbol{\epsilon}_H(k) \otimes \mathbf{I}_i \mathbf{u}_H(k) + \mathbf{F}_K \mathbf{e}_0^o(k) \right], \qquad (53)$$

where $\boldsymbol{\epsilon}_H(k)$ is the error in the estimation of the adaptive filter weight vector, i.e., $\boldsymbol{\epsilon}_H(k) = \mathbf{w}_H^o - \mathbf{w}_H(k)$. The ith block in equation (50) is given by

$$\mathbf{F}_K \mathbf{e}_0(k) \otimes \mathbf{I}_i \mathbf{u}_H^c(k) = \text{Diag}\left[\mathbf{I}_i \mathbf{u}_H^c(k)\right] \mathbf{F}_K \mathbf{e}_0(k). \qquad (54)$$

Substituting equation (53) into equation (54) yields

$$\mathbf{F}_K \mathbf{e}_0(k) \otimes \mathbf{I}_i \mathbf{u}_H^c(k) = \text{Diag}\left[\mathbf{I}_i \mathbf{u}_H^c(k)\right] \mathbf{F}_K \begin{bmatrix} \mathbf{0}_{\frac{K}{2} \times \frac{K}{2}} & \mathbf{0}_{\frac{K}{2} \times \frac{K}{2}} \\ \mathbf{0}_{\frac{K}{2} \times \frac{K}{2}} & \mathbf{I}_{\frac{K}{2} \times \frac{K}{2}} \end{bmatrix} \mathbf{F}_K^{-1} \times$$

$$\left[\text{Diag}\left[\mathbf{I}_0 \mathbf{u}_H(k)\right] \mathbf{I}_0 \boldsymbol{\epsilon}_H(k) + \mathbf{D} \sum_{i=1}^{K-1} \text{Diag}\left[\mathbf{I}_{K-i} \mathbf{u}_H(k)\right] \mathbf{I}_i \boldsymbol{\epsilon}_H(k) + \mathbf{F}_K \mathbf{e}^o(k) \right]. \qquad (55)$$

Upon defining

$$\mathbf{T}_{i,j} = \text{Diag}\left[\mathbf{I}_i \mathbf{u}_H^c(k)\right] \mathbf{L}_K \text{Diag}\left[\mathbf{I}_j \mathbf{u}_H(k)\right], \qquad (56)$$

where

$$\mathbf{L}_K = \mathbf{F}_K \begin{bmatrix} \mathbf{0}_{\frac{K}{2} \times \frac{K}{2}} & \mathbf{0}_{\frac{K}{2} \times \frac{K}{2}} \\ \mathbf{0}_{\frac{K}{2} \times \frac{K}{2}} & \mathbf{I}_{\frac{K}{2} \times \frac{K}{2}} \end{bmatrix} \mathbf{F}_K^{-1}, \tag{57}$$

the ith block of equation (50) can be written as

$$\mathbf{F}_K \mathbf{e}^o(k) \otimes \mathbf{I}_i \mathbf{u}_H^c(k) = \begin{bmatrix} \mathbf{T}_{i,0} & \mathbf{T}_{i,K-1} & \mathbf{T}_{i,K-2} & \cdots & \mathbf{T}_{i,1} \end{bmatrix} \begin{bmatrix} \mathbf{I}_0 \epsilon_H(k) \\ \mathbf{D}\mathbf{I}_1 \epsilon_H(k) \\ \mathbf{D}\mathbf{I}_2 \epsilon_H(k) \\ \vdots \\ \mathbf{D}\mathbf{I}_{K-1} \epsilon_H(k) \end{bmatrix}$$

$$+ \operatorname{Diag}\left[\mathbf{I}_i \mathbf{u}_H^c(k)\right] \mathbf{L}_K \mathbf{e}^o(k). \tag{58}$$

Using the fact that

$$\operatorname{Diag}\left[\mathbf{v}\right] \mathbf{R} \operatorname{Diag}\left[\mathbf{u}\right] = \left(\mathbf{v}\mathbf{u}^T\right) \otimes \mathbf{R}, \tag{59}$$

equation (56) can be written as

$$\mathbf{T}_{i,j} = \left(\mathbf{I}_i \mathbf{u}_H^c(k) \left(\mathbf{I}_j \mathbf{u}_H(k)\right)^T\right) \otimes \mathbf{L}_K. \tag{60}$$

Define

$$\mathbf{U}_{K^2} = \mathbf{H} \begin{bmatrix} \mathbf{I}_{\frac{K^2}{2} \times \frac{K^2}{2}} & \mathbf{0}_{\frac{K^2}{2} \times \frac{K^2}{2}} \\ \mathbf{0}_{\frac{K^2}{2} \times \frac{K^2}{2}} & \mathbf{0}_{\frac{K^2}{2} \times \frac{K^2}{2}} \end{bmatrix} \mathbf{H}^{-1}. \tag{61}$$

Then

$$\mathbf{w}_H(k+1) = \mathbf{w}_H(k)$$

$$+ \mu \, \mathbf{U}_{K^2} \mathbf{I}_K^{-1} \mathbf{T} \begin{bmatrix} \mathbf{I}_0 \epsilon_H(k) \\ \mathbf{D}\mathbf{I}_1 \epsilon_H(k) \\ \mathbf{D}\mathbf{I}_2 \epsilon_H(k) \\ \vdots \\ \mathbf{D}\mathbf{I}_{K-1} \epsilon_H(k) \end{bmatrix} + \mu \, \mathbf{U}_{K^2} \mathbf{I}_K^{-1} \begin{bmatrix} \operatorname{Diag}\left[\mathbf{I}_0 \mathbf{u}_H^c(k)\right] \\ \operatorname{Diag}\left[\mathbf{I}_1 \mathbf{u}_H^c(k)\right] \\ \operatorname{Diag}\left[\mathbf{I}_2 \mathbf{u}_H^c(k)\right] \\ \vdots \\ \operatorname{Diag}\left[\mathbf{I}_{K-1} \mathbf{u}_H^c(k)\right] \end{bmatrix} \mathbf{L}_K \mathbf{e}^o(k). \tag{62}$$

The matrix $\mathbf{T}$ can be written as

$$\mathbf{T} = \left(\mathbf{1}_{K \times K} \times \mathbf{L}_K\right) \otimes$$

$$\begin{bmatrix} \mathbf{I}_0 \mathbf{u}_H^c(k) \left[\mathbf{I}_0 \mathbf{u}_H(k)\right]^T & \mathbf{I}_0 \mathbf{u}_H^c(k) \left[\mathbf{I}_{K-1} \mathbf{u}_H(k)\right]^T & \cdots & \mathbf{I}_0 \mathbf{u}_H^c(k) \left[\mathbf{I}_1 \mathbf{u}_H(k)\right]^T \\ \mathbf{I}_1 \mathbf{u}_H^c(k) \left[\mathbf{I}_0 \mathbf{u}_H(k)\right]^T & \mathbf{I}_1 \mathbf{u}_H^c(k) \left[\mathbf{I}_{K-1} \mathbf{u}_H(k)\right]^T & \cdots & \mathbf{I}_1 \mathbf{u}_H^c(k) \left[\mathbf{I}_1 \mathbf{u}_H(k)\right]^T \\ \vdots & \vdots & \ddots & \vdots \\ \mathbf{I}_{K-1} \mathbf{u}_H^c(k) \left[\mathbf{I}_0 \mathbf{u}_H(k)\right]^T & \mathbf{I}_{K-1} \mathbf{u}_H^c(k) \left[\mathbf{I}_{K-1} \mathbf{u}_H(k)\right]^T & \cdots & \mathbf{I}_{K-1} \mathbf{u}_H^c(k) \left[\mathbf{I}_1 \mathbf{u}_H(k)\right]^T \end{bmatrix},$$

where $\times$ denotes the Kronecker product and $\mathbf{1}_{K \times K}$ is the $K \times K$ matrix with all element being equal to one. The matrix $\mathbf{T}$ can be written as

$$
\mathbf{T} = \begin{bmatrix} \mathbf{I}_0 \mathbf{u}_H^c(k) \\ \mathbf{I}_1 \mathbf{u}_H^c(k) \\ \vdots \\ \mathbf{I}_{K-2} \mathbf{u}_H^c(k) \\ \mathbf{I}_{K-1} \mathbf{u}_H^c(k) \end{bmatrix} \begin{bmatrix} \mathbf{I}_0 \mathbf{u}_H(k) \\ \mathbf{I}_{K-1} \mathbf{u}_H(k) \\ \vdots \\ \mathbf{I}_2 \mathbf{u}_H(k) \\ \mathbf{I}_1 \mathbf{u}_H(k) \end{bmatrix}^T \otimes \left(\mathbf{1}_{K \times K} \times \mathbf{L}_K \right) = \left(\mathbf{I}_K \mathbf{u}_H^c(k) \mathbf{u}_H^T(k) \mathbf{I}_K^{c\,T} \right) \otimes \left(\mathbf{1}_{K \times K} \times \mathbf{L}_K \right),
$$

where

$$
\mathbf{I}_K^c = \begin{bmatrix} \mathbf{I}_0 \\ \mathbf{I}_{K-1} \\ \vdots \\ \mathbf{I}_1 \end{bmatrix}. \tag{63}
$$

Finally, the error in the estimation of the adaptive filter is given by

$$
\epsilon_H(k+1) = \left(\mathbf{I} - \mu \mathbf{U}_{K^2} \mathbf{I}_K^{-1} \left(\mathbf{I}_K \mathbf{u}_H^c(k) \mathbf{u}_H^T(k) \mathbf{I}_K^{c\,T} \right) \otimes \left(\mathbf{1}_{K \times K} \times \mathbf{L}_K \right) \mathbf{I}_K^D \right) \epsilon_H(k)
$$

$$
- \mu \mathbf{U}_{K^2} \mathbf{I}_K^{-1} \begin{bmatrix} \mathrm{Diag}[\mathbf{I}_0 \mathbf{u}_H^c(k)] \\ \mathrm{Diag}[\mathbf{I}_1 \mathbf{u}_H^c(k)] \\ \vdots \\ \mathrm{Diag}[\mathbf{I}_{K-2} \mathbf{u}_H^c(k)] \\ \mathrm{Diag}[\mathbf{I}_{K-1} \mathbf{u}_H^c(k)] \end{bmatrix} \mathbf{L}_K \mathbf{e}^o(k), \tag{64}
$$

where

$$
\mathbf{I}_K^D = \begin{bmatrix} \mathbf{I}_0 \\ \mathbf{DI}_1 \\ \mathbf{DI}_2 \\ \vdots \\ \mathbf{DI}_{K-2} \\ \mathbf{DI}_{K-1} \end{bmatrix}. \tag{65}
$$

Therefore, the adaptive block HOT filter convergence is governed by the matrix

$$
\boldsymbol{\Psi} = \mathbf{H} \begin{bmatrix} \mathbf{I}_{\frac{K^2}{2} \times \frac{K^2}{2}} & \mathbf{0}_{\frac{K^2}{2} \times \frac{K^2}{2}} \\ \mathbf{0}_{\frac{K^2}{2} \times \frac{K^2}{2}} & \mathbf{0}_{\frac{K^2}{2} \times \frac{K^2}{2}} \end{bmatrix} \mathbf{H}^{-1} \mathbf{I}_K^{-1} \left(\mathbf{I}_K E \mathbf{u}_H^c(k) \mathbf{u}_H^T(k) \mathbf{I}_K^{cT} \right) \otimes \left(\mathbf{1}_{K \times K} \times \mathbf{L}_K \right) \mathbf{I}_K^D. \tag{66}
$$

The structure of $\mathbf{\Psi}$ is now analyzed. Using the relation between the HOT and the DFT transforms, we can write

$$\mathbf{I}_K \mathbf{u}_H^c = \begin{bmatrix} \mathbf{F}_K \mathbf{u}_0^c \\ \mathbf{F}_K \mathbf{u}_1^c \\ \vdots \\ \mathbf{F}_K \mathbf{u}_{K-2}^c \\ \mathbf{F}_K \mathbf{u}_{K-1}^c \end{bmatrix}. \tag{67}$$

It can be easily shown that

$$\mathbf{F}_K \mathbf{u}_i^c = \begin{cases} \mathbf{F}_K^H \mathbf{u}_i & \text{if } i = 0, \\ \mathbf{D}^* \mathbf{F}_K^H \mathbf{u}_{K-i} & \text{if } i \neq 0. \end{cases} \tag{68}$$

Then we have

$$\mathbf{I}_K \mathbf{u}_K^c = \begin{bmatrix} \mathbf{F}_K^H \mathbf{u}_0 \\ \mathbf{D}^* \mathbf{F}_K^H \mathbf{u}_{K-1} \\ \vdots \\ \mathbf{D}^* \mathbf{F}_K^H \mathbf{u}_2 \\ \mathbf{D}^* \mathbf{F}_K^H \mathbf{u}_1 \end{bmatrix} \tag{69}$$

and

$$\mathbf{I}_K \mathbf{u}_H^c(k) \mathbf{u}_H^T(k) \mathbf{I}_K^{cT} = \begin{bmatrix} \mathbf{F}_K^H \mathbf{u}_0 \\ \mathbf{D}^* \mathbf{F}_K^H \mathbf{u}_{K-1} \\ \vdots \\ \mathbf{D}^* \mathbf{F}_K^H \mathbf{u}_2 \\ \mathbf{D}^* \mathbf{F}_K^H \mathbf{u}_1 \end{bmatrix} \begin{bmatrix} \mathbf{F}_K \mathbf{u}_0 \\ \mathbf{F}_K \mathbf{u}_{K-1} \\ \vdots \\ \mathbf{F}_K \mathbf{u}_2 \\ \mathbf{F}_K \mathbf{u}_1 \end{bmatrix}^T. \tag{70}$$

Taking the expectation of equation (70) yields

$$\mathbf{I}_K E \mathbf{u}_H^c(k) \mathbf{u}_H^T(k) \mathbf{I}_K^{cT} = \begin{bmatrix} \mathbf{F}_K^H E \mathbf{u}_0 \mathbf{u}_0^T \mathbf{F}_K & \mathbf{F}_K^H E \mathbf{u}_0 \mathbf{u}_{K-1}^T \mathbf{F}_K & \cdots & \mathbf{F}_K^H E \mathbf{u}_0 \mathbf{u}_1^T \mathbf{F}_K \\ \mathbf{D}^* \mathbf{F}_K^H E \mathbf{u}_{K-1} \mathbf{u}_0^T \mathbf{F}_K & \mathbf{D}^* \mathbf{F}_K^H E \mathbf{u}_{K-1} \mathbf{u}_{K-1}^T \mathbf{F}_K & \cdots & \mathbf{D}^* \mathbf{F}_K^H E \mathbf{u}_{K-1} \mathbf{u}_1^T \mathbf{F}_K \\ \vdots & \vdots & \ddots & \vdots \\ \mathbf{D}^* \mathbf{F}_K^H E \mathbf{u}_1 \mathbf{u}_0^T \mathbf{F}_K & \mathbf{D}^* \mathbf{F}_K^H E \mathbf{u}_1 \mathbf{u}_{K-1}^T \mathbf{F}_K & \cdots & \mathbf{D}^* \mathbf{F}_K^H E \mathbf{u}_1 \mathbf{u}_1^T \mathbf{F}_K \end{bmatrix}.$$

Each block in the above equation is an autocorrelation matrix that is asymptotically diagonalized by the DFT matrix. Each block will be also pointwise multiplied by $\mathbf{L}_K$. Three-dimensional representations of $\mathbf{L}_K$ for $K = 16$ and $K = 32$ are shown in Figures 5 and 6, respectively. The diagonal elements of $\mathbf{L}_K$ are much higher than the off diagonal elements. Therefore, pointwise multiplying each block in the previous equation with $\mathbf{L}_K$ makes it more diagonal. If each block is perfectly diagonal, then $\mathbf{I}_K \left(\mathbf{I}_K E \mathbf{u}_H^c(k) \mathbf{u}_H^T(k) \mathbf{I}_K^{cT} \right) \otimes \left(\mathbf{1}_{K \times K} \times \mathbf{L}_K \right) \mathbf{I}_K^D$ will be block diagonal. Asymptotically the HOT block LMS adaptive filter transforms the K^2 modes into K decoupled sets of modes.

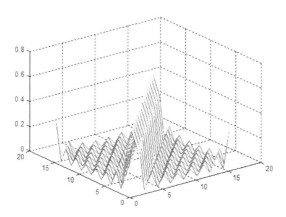

Figure 5. Three-dimensional representation of $\mathbf{L}_{16}$.

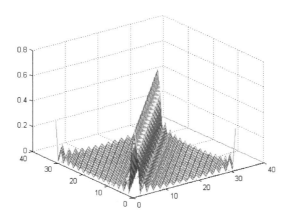

Figure 6. Three-dimensional representation of $\mathbf{L}_{32}$.

8. Conclusions

The "HOT convolution," a relation between the HOT of two signals and their circular convolution was derived. The result was used to develop a fast block LMS adaptive filter called the HOT block LMS adaptive filter. The HOT block LMS adaptive filter assumes that the filter and block lengths are the same. This filter requires slightly less than half of the multiplications that are required for the DFT block LMS adaptive filter. The reduction in the computational complexity of the HOT block LMS comes from using only one polyphase component of the filter error used to update the filter weights. Convergence analysis of the HOT block LMS algorithm showed that the average time constant is the same as that of the DFT block LMS algorithm and that the misadjustment is K times greater than that of the DFT block LMS algorithm. The HOT block LMS adaptive filter transforms the K^2 modes into K decoupled sets of modes.

Author details

Osama Alkhouli[1,*],
Victor DeBrunner[2] and Joseph Havlicek[3]

* Address all correspondence to: osama_k@ou.edu

1 Caterpillar Inc., Illinois, USA
2 University of Oklahoma, School of Electrical and Computer Engineering, Oklahoma, USA
3 Florida State University, Electrical and Computer Engineering Department, Florida, USA

References

[1] I. I. Hirschman, "A note on entropy," *Amer. J. Math.*, vol. 79, pp. 152-156, 1957.

[2] H T. Przebinda, V. DeBrunner, and M. Özaydin, "The optimal transform for the discrete Hirschman uncertainty principle," *IEEE Trans. Infor. Theory*, pp. 2086-2090, Jul 2001.

[3] V. DeBrunner, M. Özaydin, and T. Przebinda, "Resolution in time-frequency," *IEEE Trans. ASSP*, pp. 783-788, Mar 1999.

[4] V. DeBrunner, J. Havlicek, T. Przebinda, and M. Özaydin, "Entropy-based uncertainty measures for $L^2(R)^n$, $\ell^2(Z)$, and $\ell^2(Z/NZ)$ with a Hirschman optimal transform for $\ell^2(Z/NZ)$," *IEEE Trans. ASSP*, pp. 2690-2696, August 2005.

[5] V. DeBrunner and E. Matusiak, "An algorithm to reduce the complexity required to convolve finite length sequences using the Hirschman optimal transform (HOT)," *ICASSP 2003*, Hong Kong, China, pp. II-577-580, Apr 2003.

[6] G. Clark, S. Mitra, and S Parker, "Block implementation of adaptive digital filters," *IEEE Trans. ASSP*, pp. 744-752, Jun 1981.

[7] E. R. Ferrara, "Fast implementation of LMS adaptive filters," *IEEE Trans. ASSP*, vol. ASSP-28, NO. 4, Aug 1980.

[8] J. Benesty and P. Duhamel, "A fast exact least mean square adaptive algorithm," *IEEE Trans. ASSP*, pp. 2904-2920, Dec 1992.

[9] D. Mansour and A. H. Gray, "Unconstrained frequency-domain adaptive filter," *IEEE Trans. ASSP*, pp. 726-734, Oct 1982.

[10] Simon Haykin, *Adaptive Filter Theory*. Prentice Hall information and system sciences series, Fourth edition, 2002.

Chapter 2

Applications of a Combination of Two Adaptive Filters

Tõnu Trump

Additional information is available at the end of the chapter

1. Introduction

Designing a Least Mean Square (LMS) family adaptive algorithm includes solving the well-known trade-off between the initial convergence speed and the mean-square error in steady state according to the requirements of the application at hands. The trade-off is controlled by the step-size parameter of the algorithm. Large step size leads to a fast initial convergence but the algorithm also exhibits a large mean-square error in the steady state and in contrary, small step size slows down the convergence but results in a small steady state error [9,17]. In several applications it is, however, eligible to have both and hence it would be very desirable to be able to design algorithms that can overcome the named trade-off.

Variable step size adaptive schemes offer a potential solution allowing to achieve both fast initial convergence and low steady state misadjustment [1, 8, 12, 15, 18]. How successful these schemes are depends on how well the algorithm is able to estimate the distance of the adaptive filter weights from the optimal solution. The variable step size algorithms use different criteria for calculating the proper step size at any given time instance. For example the algorithm proposed in [15] changes the time-varying convergence parameters in such a way that the change is proportional to the negative of gradient of the squared estimation error with respect to the convergence parameter. Squared instantaneous errors have been used in [12] and the squared autocorrelation of errors at adjacent time instances in [1] to modify the step size. In reference [18] the norm of projected weight error vector is used as a criterion to determine how close the adaptive filter is to its optimum performance.

More recently there has been an interest in a combination scheme that is able to optimize the trade-off between convergence speed and steady state error [14]. The scheme consists of two adaptive filters that are simultaneously applied to the same inputs as depicted in Figure 1. One of the filters has a large step size allowing fast convergence and the other one has a small step size for a small steady state error. The outputs of the filters are combined through a mixing parameter λ. The performance of this scheme has been studied for some parameter update schemes [2, 6, 19]. The reference [2] uses convex combination i.e. λ is constrained to lie between 0 and 1. The reference [19] presents a transient analysis of a slightly modified version of this scheme. The parameter λ is in those papers found using an LMS type adaptive scheme and computing the sigmoidal function of the result. The reference [6] takes another approach computing the mixing parameter using an affine combination. This paper uses the ratio of time averages of the instantaneous errors of the filters. The error function of the ratio is then computed to obtain λ.

In [13] a convex combination of two adaptive filters with different adaptation schemes has been investigated with the aim to improve the steady state characteristics. One of the adaptive filters in that paper uses LMS algorithm and the other one Generalized Normalized Gradient Decent algorithm. The combination parameter λ is computed using stochastic gradient adaptation. In [24] the convex combination of two adaptive filters is applied in a variable filter length scheme to gain improvements in low SNR conditions. In [11] the combination has been used to join two affine projection filters with different regularization parameters. The work [7] uses the combination on parallel binary structured LMS algorithms. These three works use the LMS like scheme of [5] to compute λ.

It should be noted that schemes involving two filters have been proposed earlier [3, 16]. However, in those early schemes only one of the filters have been adaptive while the other one has used fixed filter weights. Updating of the fixed filter has been accomplished by copying of all the coefficients from the adaptive filter, when the adaptive filter has been performing better than the fixed one.

In this Chapter we compute the mixing parameter λ from output signals of the individual filters. The way of calculating the mixing parameter is optimal in the sense that it results from minimization of the mean-squared error of the combined filter. The scheme was independently proposed in [21] and [4]. In [23], the output signal based combination was used in adaptive line enhancer and in [22] it was used in the system identification application.

We will investigate three applications of the combination: system investigation, adaptive beamforming and adaptive line enhancer. We describe each of the applications in detail and present a proper analysis.

We will assume throughout the Chapter that the signals are complex-valued and that the combination scheme uses two LMS adaptive filters. The italic, bold face lower case and bold face upper case letters will be used for scalars, column vectors and matrices respectively. The superscript T denotes transposition and the superscript H Hermitian transposition of a matrix. The operator $E[\cdot]$ denotes mathematical expectation, $Re\{\cdot\}$ is the real part of a complex variable and $Tr[\cdot]$ stands for the trace of a matrix.

2. Combination of Two Adaptive Filters

Let us consider two adaptive filters, as shown in Figure 1, each of them updated using the LMS adaptation rule

$$e_i(n) = d(n) - \mathbf{w}_i^H(n-1)\mathbf{x}(n), \tag{1}$$

$$\mathbf{w}_i(n) = \mathbf{w}_i(n-1) + \mu_i e_i^*(n)\mathbf{x}(n). \tag{2}$$

In the above $w_i(n)$ is the N vector of coefficients of the i-th adaptive filter, with $i = 1,2$ and $x(n)$ is the known N input vector, common for both of the adaptive filters. The input process is assumed to be a zero mean wide sense stationary Gaussian process. μ_i is the step size of i-th adaptive filter. We assume without loss of generality that $\mu_1 > \mu_2$. The case $\mu_1 = \mu_2$ is not interesting as in this case the two filters remain equal and the combination renders to a single filter.

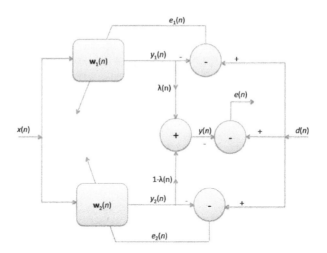

Figure 1. The combined adaptive filter.

The desired signal in 1 can be expressed as

$$d(n) = \mathbf{w}_o^H \mathbf{x}(n) + \zeta(n)., \tag{3}$$

where the vector w_o is the optimal Wiener filter coefficient vector for the problem at hands and the process $\zeta(n)$ is the irreducible error that is statistically independent of all the other signals.

The outputs of the two adaptive filters are combined according to

$$y(n) = \lambda(n)y_1(n) + [1 - \lambda(n)]y_2(n),$$ (4)

where $y_i(n) = \mathbf{w}_i^H(n-1)\mathbf{x}(n)$ and the mixing parameter $\lambda(n)$ can be any real number.

We define the *a priori* system error signal as difference between the output signal of the optimal Wiener filter at time n, given by $y_o(n) = \mathbf{w}_o^H \mathbf{x}(n) = d(n) - \zeta(n)$, and the output signal of our adaptive scheme $y(n)$

$$e_a(n) = y_o(n) - \lambda(n)y_1(n) - (1 - \lambda(n))y_2(n).$$ (5)

Let us now find $\lambda(n)$ by minimizing the mean square of the *a priori* system error. The derivative of $E[\,|\,e_a(n)\,|^2\,]$ with respect to $\lambda(n)$ reads

$$\frac{\partial E[\,|\,e_a(n)\,|^2\,]}{\partial \lambda(n)} = 2E[\operatorname{Re}\{(y_o(n) - y_2(n))(y_2(n) - y_1(n))^*\} + \lambda(n)\,|\,(y_2(n) - y_1(n))\,|^2\,].$$ (6)

Setting the derivative to zero results in

$$\lambda(n) = \frac{E[\operatorname{Re}\{(d(n) - y_2(n))(y_1(n) - y_2(n))^*\}]}{E[\,|\,(y_1(n) - y_2(n))\,|^2\,]},$$ (7)

where we have replaced the Wiener filter output signal $y_o(n)$ by its observable noisy version $d(n)$. Note however, that because the input signal $x(n)$ and irreducible error $\zeta(n)$ are independent random processes, this can be done without introducing any error into our calculations. The denominator of equation (7) comprises expectation of the squared difference of the two filter output signals. This quantity can be very small or even zero, particularly in the beginning of adaptation if the two step sizes are close to each other. Correspondingly λ computed directly from (7) may be large. To avoid this from happening we add a small regularization constant to the denominator of (7). The constant should be selected small compared to $E[\mathbf{x}^T(n)\mathbf{x}(n)]$ but large enough to prevent division by zero in given arithmetic.

3. System Identification

In several areas it is essential to build a mathematical model of some phenomenon or system. In this class of applications, the adaptive filter can be used to find a best fit of a linear model to an unknown plant. The plant and the adaptive filter are driven by the same known input signal and the plant output provides the desired signal of the adaptive filter. The plant can be dynamic and in this case we have a time varying model. The system identification configuration is depicted in Figure 2. As before $x(n)$ is the input signal, $v(n)$ is the measurement noise, $y(n)$ is the adaptive filter output signal and $e(n)$ is the error signal. The desired signal is $d(n) = \mathbf{w}_o^H \mathbf{x}(n) + \zeta(n)$, where w_o is the vector of Wiener filter coefficients and the irreducible error $\zeta(n)$ consists of the measurement noise $v(n)$ together with the effects of the plant that can not be explained with a length N linear model. The result of pure system identification problem is the vector of adaptive filter coefficients.

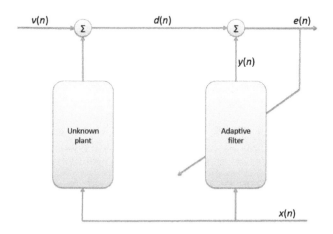

Figure 2. Block diagram of generelized sidelobe canceller.

The same basic configuration is also used to solve the echo and noise cancellation problems. In echo cancellation the unknown plant is the echo path either electrical or acoustical and the input signal $x(n)$ is the speech signal of one of the parties of telephone conversation. Speech of the other party is contained in the signal $v(n)$. The objective is to cancel the components of the desired signal that are due to the input $x(n)$.

In noise cancellation problems the signal $v(n)$ is the primary microphone signal containing noise and the signal to be cleaned. The input signal $x(n)$ is formed by the reference microphones. The reference signals are supposed to be correlated with the noise in the primary signal but not with the useful signal. The objective here is to suppress the noise and clean the signal of interest i.e. $v(n)$.

In here we are going to use the combination of two adaptive filters described in the previous Section to solve the system identification problem.

3.2 Excess Mean Square Error

In this section we are interested in finding expressions that characterize transient performance of the combined algorithm i.e. we intend to derive formulae that predict entire course of adaptation of the algorithm. Before we can proceed we need, however, to introduce some notations.

First let us denote the weight error vector of i-th filter as

$$\tilde{\mathbf{w}}_i(n) = \mathbf{w}_o - \mathbf{w}_i(n). \tag{8}$$

Then the equivalent weight error vector of the combined adaptive filter will be

$$\tilde{\mathbf{w}}(n) = \lambda \tilde{\mathbf{w}}_1(n) + (1-\lambda)\tilde{\mathbf{w}}_2(n). \tag{9}$$

The mean square deviation of the combined filter $MSD = E[\tilde{\mathbf{w}}^H(n)\tilde{\mathbf{w}}(n)]$ is given by

$$MSD = \lambda^2 E[\tilde{\mathbf{w}}_1^H(n)\tilde{\mathbf{w}}_1(n)] + 2\lambda(1-\lambda)\text{Re}\{E[\tilde{\mathbf{w}}_2^H(n)\tilde{\mathbf{w}}_1(n)]\} + (1-\lambda)^2 E[\tilde{\mathbf{w}}_2^H(n)\tilde{\mathbf{w}}_2(n)]. \tag{10}$$

The *a priori* estimation error of an individual filter is defined as

$$e_{i,a}(n) = \tilde{\mathbf{w}}_i^H(n-1)\mathbf{x}(n). \tag{11}$$

It follows from (5) that we can express the *a priori* error of the combination as

$$e_a(n) = \lambda(n)e_{1,a}(n) + (1-\lambda(n))e_{2,a}(n) \tag{12}$$

and because $\lambda(n)$ is according to (7) a ratio of mathematical expectations and, hence, deterministic, we have for the excess mean square error of the combination, $EMSE(n) = E[|e_a(n)|^2]$,

$$E[|e_a(n)|^2] = \lambda^2 E[|e_{1,a}(n)|^2] + 2\lambda(1-\lambda)E[\text{Re}\{e_{1,a}(n)e_{2,a}^*(n)\}] + (1-\lambda)^2 E[|e_{2,a}(n)|^2]. \tag{13}$$

As $e_{i,a}(n) = \tilde{\mathbf{w}}_i^H(n-1)\mathbf{x}(n)$, the expression of the excess mean square error becomes

$$E[|e_a(n)|^2] = \lambda^2 E[\tilde{\mathbf{w}}_1^H \mathbf{x}\mathbf{x}^H \tilde{\mathbf{w}}_1] + 2\lambda(1-\lambda)E[\text{Re}\{\tilde{\mathbf{w}}_1^H \mathbf{x}\mathbf{x}^H \tilde{\mathbf{w}}_2\}] + (1-\lambda)^2 E[\tilde{\mathbf{w}}_2^H \mathbf{x}\mathbf{x}^H \tilde{\mathbf{w}}_2]. \tag{14}$$

In what follows we often drop the explicit time index n as we have done in (14), if it is not necessary to avoid a confusion.

Noting that $y_i(n) = \mathbf{w}_i^H(n-1)\mathbf{x}(n)$, we can rewrite the expression for $\lambda(n)$ in (7) as

$$\lambda(n) = \frac{E[\tilde{\mathbf{w}}_2^H \mathbf{x}\mathbf{x}^H \tilde{\mathbf{w}}_2] - E[\text{Re}\{\tilde{\mathbf{w}}_2^H \mathbf{x}\mathbf{x}^H \tilde{\mathbf{w}}_1\}]}{E[\tilde{\mathbf{w}}_1^H \mathbf{x}\mathbf{x}^H \tilde{\mathbf{w}}_1] - 2E[\text{Re}\{\tilde{\mathbf{w}}_1^H \mathbf{x}\mathbf{x}^H \tilde{\mathbf{w}}_2\}] + E[\tilde{\mathbf{w}}_2^H \mathbf{x}\mathbf{x}^H \tilde{\mathbf{w}}_2]}. \tag{15}$$

We thus need to investigate the evolution of the individual terms of the type $EMSE_{k,l} = E[\tilde{\mathbf{w}}_k^H(n-1)\mathbf{x}(n)\mathbf{x}^H(n)\tilde{\mathbf{w}}_l(n-1)]$ in order to reveal the time evolution of $EMSE(n)$ and $\lambda(n)$. To do so we, however, concentrate first on the mean square deviation defined in (10).

Reformulation the relation (1) as

$$e_i(n) = d(n) - \mathbf{w}_i^H(n-1)\mathbf{x}(n) = e_o(n) + \tilde{\mathbf{w}}_i^H(n-1)\mathbf{x}(n) \tag{16}$$

and subtracting (2) from w_o we have

$$\tilde{\mathbf{w}}_i(n) = (\mathbf{I} - \mu_i \mathbf{x}\mathbf{x}^H)\tilde{\mathbf{w}}_i(n-1) - \mu_i \mathbf{x}e_o^*(n). \tag{17}$$

We next approximate the outer product of input signal vectors by its correlation matrix $\mathbf{x}\mathbf{x}^H \approx \mathbf{R}_\mathbf{x}$. The approximation is justified by the fact that with small step size the weight error update of the LMS algorithm (17) behaves like a low pass filter with a low cutoff frequency. With this approximations we have

$$\tilde{\mathbf{w}}_i(n) \approx (\mathbf{I} - \mu_i \mathbf{R}_\mathbf{x})\tilde{\mathbf{w}}_i(n-1) - \mu_i \mathbf{x}e_o^*(n). \tag{18}$$

This means in fact that we apply the small step size theory [9] even if the assumption of small step size is not really true for the fast adapting filter. In our simulation study we will see, however, that the assumption works in practice rather well.

Let us now define the eigendecomposition of the correlation matrix as

$$\mathbf{Q}^H \mathbf{R}_x \mathbf{Q} = \Omega, \tag{19}$$

where $\mathbf{Q}$ is a unitary matrix whose columns are the orthogonal eigenvectors of $\mathbf{R}_x$ and Ω is a diagonal matrix having eigenvalues associated with the corresponding eigenvectors on its main diagonal. We also define the transformed weight error vector as

$$\mathbf{v}_i(n) = \mathbf{Q}^H \tilde{\mathbf{w}}_i(n) \tag{20}$$

and the transformed last term of equation (18) as

$$\mathbf{p}_i(n) = \mu_i \mathbf{Q}^H \mathbf{x}e_o^*(n). \tag{21}$$

Then we can rewrite the equation (18) after multiplying both sides by $\mathbf{Q}^H$ from the left as

$$\mathbf{v}_i(n) = (\mathbf{I} - \mu_i \mathbf{\Omega}) \mathbf{v}_i(n-1) - \mathbf{p}_i(n). \tag{22}$$

We note that the mean of p_i is zero by the orthogonality theorem and the crosscorrelation matrix of p_k and p_l equals

$$E[\mathbf{p}_k \mathbf{p}_l^H] = \mu_k \mu_l \mathbf{Q}^H E[xe_o^*(n) e_o(n) \mathbf{x}^H] \mathbf{Q}. \tag{23}$$

We now invoke the Gaussian moment factoring theorem to write

$$E[\mathbf{x}e_o^*(n) e_o(n) \mathbf{x}^H] = E[\mathbf{x}e_o^*(n)] E[e_o(n) \mathbf{x}^H] + E[\mathbf{x}\mathbf{x}^H] E[|e_o|^2]. \tag{24}$$

The first term in the above is zero due to the principle of orthogonality and the second term equals $\mathbf{R}J_{min}$, where $J_{min} = E[|e_o|^2]$ is the minimum mean square error produced by the corresponding Wiener filter. Hence we are left with

$$E[\mathbf{p}_k \mathbf{p}_l^H] = \mu_k \mu_l J_{min} \mathbf{\Omega}. \tag{25}$$

As the matrices I and Ω in (22) are both diagonal, it follows that the m-th element of vector $v_i(n)$ is given by

$$\begin{aligned} v_{i,m}(n) &= (1 - \mu_i \omega_m) v_{i,m}(n-1) - p_{i,m}(n) \\ &= (1 - \mu_i \omega_m)^n v_m(0) + \sum_{i=0}^{n-1} (1 - \mu_i \omega_m)^{n-1-i} p_{i,m}(i), \end{aligned} \tag{26}$$

where ω_m is the m-th eigenvalue of R_x and $v_{i,m}$ and $p_{i,m}$ are the m-th components of the vectors v_i and p_i respectively.

We immediately see that the mean value of $v_{i,m}(n)$ equals

$$E[v_{i,m}(n)] = (1 - \mu_i \omega_m)^n v_m(0) \tag{27}$$

as the vector p_i has zero mean.

To proceed with our development for the combination of two LMS filters we note that we can express the MSD and its individual components in (10) through the transformed weight error vectors as

$$\begin{aligned} E[\tilde{\mathbf{w}}_k^H(n) \tilde{\mathbf{w}}_l(n)] &= E[\mathbf{v}_k^H(n) \mathbf{v}_l(n)] \\ &= \sum_{m=0}^{N-1} E[v_{k,m}(n) v_{l,m}^*(n)] \end{aligned} \tag{28}$$

so we also need to find the auto- and cross correlations of v.

Let us concentrate on the m-th component in the sum above corresponding to the cross term. The expressions for the component filters follow as special cases. Substituting (26) into the expression of m-th component of MSD above, taking the mathematical expectation and noting that the vector p is independent of $v(0)$ results in

$$
\begin{aligned}
E[v_{k,m}(n)v_{l,m}^*(n)] = & \, E\left[(1-\mu_k\omega_m)^n v_k(0)(1-\mu_l\omega_m)^n v_l^*(0)\right] \\
& + E\left[\sum_{i=0}^{n-1}\sum_{j=0}^{n-1}(1-\mu_k\omega_m)^{n-1-i}(1-\mu_l\omega_m)^{n-1-j}p_{k,m}(i)p_{l,m}^*(j)\right].
\end{aligned}
\tag{29}
$$

We now note that most likely the two component filters are initialized to the same value

$$
[v_{k,m}(0)=v_{l,m}(0)=v_m(0)]
$$

and that

$$
E[p_{k,m}(i)p_{l,m}^*(j)] = \begin{cases} \mu_k\mu_l\omega_m J_{min}, & i=j \\ 0, & \text{otherwise} \end{cases}.
\tag{30}
$$

We then have for the m-th component of MSD

$$
\begin{aligned}
E[v_{k,m}(n)v_{l,m}^*(n)] = & \, (1-\mu_k\omega_m)^n(1-\mu_l\omega_m)^n \mid v_m(0)\mid^2 \\
& + \mu_k\mu_l\omega_m J_{min}(1-\mu_k\omega_m)^{n-1}(1-\mu_l\omega_m)^{n-1} \\
& \cdot \sum_{i=0}^{n-1}(1-\mu_k\omega_m)^{-i}(1-\mu_l\omega_m)^{-i}.
\end{aligned}
\tag{31}
$$

The sum over i in the above equation can be recognized as a geometric series with n terms. The first term is equal to 1 and the geometric ratio equals $(1-\mu_k\omega_m)^{-1}(1-\mu_l\omega_m)^{-1}$. Hence we have

$$
\begin{aligned}
\sum_{i=0}^{n-1}(1-\mu_k\omega_m)^{-i}(1-\mu_l\omega_m)^{-i} & = \frac{1-\left[(1-\mu_k\lambda_m)^{-1}(1-\mu_l\lambda_m)^{-1}\right]^n}{1-(1-\mu_k\lambda_m)^{-1}(1-\mu_l\lambda_m)^{-1}} \\
& = \frac{(1-\mu_k\omega_m)(1-\mu_l\omega_m)}{\mu_k\mu_l\omega_m^2-\mu_k\omega_m-\mu_l\omega_m} - \frac{(1-\mu_k\omega_m)^{-n+1}(1-\mu_l\omega_m)^{-n+1}}{\mu_k\mu_l\omega_m^2-\mu_k\omega_m-\mu_l\omega_m}.
\end{aligned}
\tag{32}
$$

After substitution of the above into (31) and simplification we are left with

$$E[v_{k,m}(n)v_{l,m}^*(n)] = \frac{(1-\mu_k\omega_m)^n(1-\mu_l\omega_m)^n}{} \left[\mid v_m(0)\mid^2 + \frac{J_{min}}{\omega_m^2 - \frac{\omega_m}{\mu_l} - \frac{\omega_m}{\mu_k}} \right]$$
$$- \frac{J_{min}}{\omega_m^2 - \frac{\omega_m}{\mu_l} - \frac{\omega_m}{\mu_k}},$$

(33)

which is our result for a single entry to the MSD crossterm vector. It is easy to see that for the terms involving a single filter we get an expressions that coincide with the one available in the literature [9].

Let us now focus on the cross term

$$EMSE_{kl} = E\big[\tilde{\mathbf{w}}_k^H(n-1)\mathbf{x}(n)\mathbf{x}^H(n)\tilde{\mathbf{w}}_l(n-1)\big],$$

appearing in the EMSE equation (14). Due to the independence assumption we can rewrite this using the properties of trace operator as

$$\begin{aligned} EMSE_{kl} &= E\big[\tilde{\mathbf{w}}_k^H(n-1)\mathbf{R}_x\tilde{\mathbf{w}}_l(n-1)\big] \\ &= Tr\big\{E\big[\mathbf{R}_x\tilde{\mathbf{w}}_l(n-1)\tilde{\mathbf{w}}_k^H(n-1)\big]\big\} \\ &= Tr\big\{\mathbf{R}_x E\big[\tilde{\mathbf{w}}_l(n-1)\tilde{\mathbf{w}}_k^H(n-1)\big]\big\}. \end{aligned}$$

(34)

Let us now recall that according to (20) for any of the filters $\tilde{\mathbf{w}}_i(n)=\mathbf{Q}\mathbf{v}_i(n)$ so that we are justified to write

$$\begin{aligned} EMSE_{kl} &= Tr\big\{\mathbf{R}_x E\big[\mathbf{Q}\mathbf{v}_l(n-1)\mathbf{v}_k^H(n-1)\mathbf{Q}^H\big]\big\} \\ &= Tr\big\{E\big[\mathbf{v}_k^H(n-1)\mathbf{Q}^H\mathbf{R}_x\mathbf{Q}\mathbf{v}_l(n-1)\big]\big\} \\ &= Tr\big\{E\big[\mathbf{v}_k^H(n-1)\mathbf{\Omega}\mathbf{v}_l(n-1)\big]\big\} \\ &= \sum_{i=0}^{N-1}\omega_i E\big[v_{k,i}^*(n-1)v_{l,i}(n-1)\big]. \end{aligned}$$

(35)

The EMSE of the combined filter can now be computed as

$$EMSE = \sum_{i=0}^{N-1}\omega_i E\big[\mid \lambda(n)v_{k,i}(n-1) + (1-\lambda(n))v_{l,i}(n-1)\mid^2\big],$$

(36)

where the components of type $E[v_{k,i}(n-1)v_{l,i}(n-1)]$ are given by (33). To compute $\lambda(n)$ we use (15) substituting (35) for its individual components.

4. Adaptive Sensor Array

In this Chapter we describe how to use the combination of two adaptive filters in an adaptive beamformer. The beamformer we employ here is often termed as Generalized Sidelobe Canceller [9].

Let ϕ denote the the angle of incidence of a planar wave impinging a linear sensor array, measured with respect to the normal to the array. The electrical angle θ is related to the incidence angle as

$$\theta = \frac{2\pi\delta}{\lambda}\sin\phi, \tag{37}$$

where λ is the wavelength of the incident wave and δ is the spacing between adjacent sensors of the linear array.

Suppose that the signal impinging the array of $M=N+1$ sensors is given by

$$\mathbf{u}(n) = \mathbf{A}(\Theta)\mathbf{s}(n) + \mathbf{v}(n), \tag{38}$$

where $s(n)$ is the vector of emitter signals, Θ is a collection of directions of arrivals, $A(\Theta)$ is the array steering matrix with its columns $a(\theta)$ defined as responses toward the individual sources $s(n)$ and $v(n)$ is a vector of additive circularly symmetric Gaussian noise. The M vectors

$$\mathbf{a}(\theta) = [1, e^{j\theta}, ..., e^{j(M-1)\theta}]^T \tag{39}$$

are called the steering vectors of the respective sources. We assume that the source of interest is located at the electrical angle θ_0.

The block diagram of the Generalized Sidelobe Canceller is shown in Figure 3. The structure consists of two branches. The upper branch is the steering branch, that directs its beam toward the desired source. The lower branch is the blocking branch that blocks the signals impinging at the array from the direction of the desired source and includes an adaptive algorithm that minimizes the mean square error between the output signals of the branches.

The weights in steering branch $\mathbf{w}_s$ are selected from the condition

$$\mathbf{w}_s^H \mathbf{a}(\theta_0) = g \tag{40}$$

i.e. we require the response in the direction of the source of interest θ_0 to equal a constant g. Common choices for g are $g=M$ and $g=1$. Here we have used $g=M$.

The signal at the output of the upper branch is given by

$$d(n) = \mathbf{w}_s^H \mathbf{u}(n).$$ (41)

In the lower branch we have a blocking matrix, that will block any signal coming from the direction θ_0. The columns of the $M \times M\text{-}1$ blocking matrix C_b are defined as being the orthogonal complement of the steering vector $a(\theta_0)$ in the upper branch

$$\mathbf{a}^H(\theta_0)\mathbf{C}_b = 0.$$ (42)

The vector valued signal $x(n)$ at the output of the blocking matrix is formed as

$$\mathbf{x}(n) = \mathbf{C}_b^H \mathbf{u}(n).$$ (43)

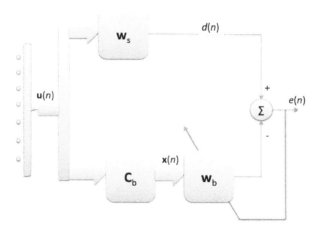

Figure 3. Block diagram of generelized sidelobe canceller.

The output of the algorithm is

$$e(n) = d(n) - \mathbf{w}_b^H(n)\mathbf{x}(n).$$ (44)

The signals $x(n)$ and $d(n)$ can be used as the input and desired signals respectively in an adaptive algorithm to select the blocking weights w_b. In this Chapter we use the combination of two adaptive filters that gives us fast initial convergence and low steady state misadjustment at the same time.

4.1 Signal to Interference and Noise Ratio

The EMSE of the adaptive algorithm can be analysed as it is done in Section 3.1. In this application we are also interested in signal to interference and noise ratio (SINR) at the array output. To evaluate this we first note that the power that signal of interest generates at the array output is according to (40)

$$P_s = \mathbf{w}_s^H \mathbf{a}(\theta_0) \sigma_{s0}^2 \mathbf{a}^H(\theta_0) \mathbf{w}_s = \mid g \mid^2 \sigma_{s0}^2, \tag{45}$$

where σ_{s0}^2 is the variance of the useful signal arriving from the angle θ_0.

To find the interference and noise power we first define the reduced signal vector $\breve{\mathbf{s}}$ and a reduced DOA collection $\breve{\Theta}$ where we have left out the signal of interest and the steering vector corresponding to the useful signal but kept all the interferers and the interference steering vectors. The corresponding array steering matrix is $\breve{\mathbf{A}}(\breve{\Theta})$.

The correlation matrix of interference and noise in the signal $x(n)$, which is the input signal to our adaptive scheme, is then given by

$$\breve{\mathbf{R}}_x = \mathbf{C}_b^H \breve{\mathbf{A}}(\breve{\Theta}) E[\breve{\mathbf{s}}\breve{\mathbf{s}}^H] \breve{\mathbf{A}}^H(\breve{\Theta}) \mathbf{C}_b + \mathbf{C}_s^H \mathbf{C}_b \sigma_v^2, \tag{46}$$

where $sigma_v^2$ is the noise variance, the first component in the summation is due to the interfering sources and the second component is due to the noise.

It follows from the standard Wiener filtering theory that the minimum interference and noise power at the array output is given by

$$\breve{J}_{min} = \sigma_{int,v}^2 - \breve{\mathbf{p}}^H \breve{\mathbf{R}}^{-1} \breve{\mathbf{p}}, \tag{47}$$

where the desired signal variance excluding the signal from the source of interest is

$$\sigma_{int,v}^2 = \mathbf{w}_s^H \breve{\mathbf{A}} \breve{\mathbf{R}} \breve{\mathbf{A}}^H \mathbf{w}_s + \sigma_v^2 \mathbf{w}_s^H \mathbf{w}_s \tag{48}$$

and the crosscorrelation vector between the adaptive filter input signal and desired signal excluding the signal from source of interest is

$$\breve{\mathbf{p}} = \mathbf{C}_b^H \breve{\mathbf{A}}(\breve{\Theta}) E[\breve{\mathbf{s}}\breve{\mathbf{s}}^H] \breve{\mathbf{A}}^H(\breve{\Theta}) \mathbf{w}_s + \sigma_v^2 \mathbf{C}_b^H \breve{\mathbf{A}}(\breve{\Theta}) \mathbf{w}_s. \tag{49}$$

We can now find the eigendecomposition of $\breve{\mathbf{R}}_x$ and use the resulting eigenvalues in (35) and (36) to find the excess mean square error due to interference and noise only $EMSE_{int,v}$. The error power can be computed as minimum interference and noise power at the array output plus excess mean square error due to interference and noise only

$$P_{v,int} = \breve{J}_{min} + EMSE_{int,v}(n) \tag{50}$$

and the signal to noise ratio is thus given by

$$SNR(n) = \frac{P_s}{P_{v,int}(n)}. \tag{51}$$

5. Adaptive Line Enhancer

Adaptive line enhancer is a device that is able to separate the input into two components. One of them consists mostly of the narrow-band signals that are present at the input and the other one consists mostly of the broadband noise. In the context of this paper the signal is considered to be of narrow band if its bandwidth is small as compared to the sampling frequency of the system.

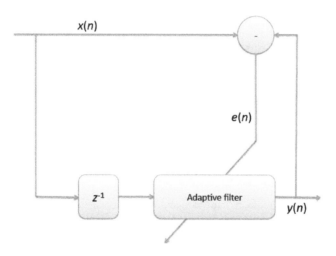

Figure 4. The adaptive line enhancer.

We assume that the broadband noise is zero mean, white and Gaussian and that the narrow band component is centred. One is usually interested in the narrow band components and the device is often used to clean narrow band signals from noise before any further processing. The line enhancer is shown in Figure 4. Note that the input signal to the adaptive filter of the line enhancer is delayed by Δ sample times and the input vector is thus $x(n-\Delta)$. The desired signal is $d(n) = x(n)$. The line enhancer is in fact a Δ step predictor. The device is able to predict the narrow band components that have long correlation times but it cannot predict the white noise and hence, only a prediction of narrow band components appears in the filter output signal $y(n)$. The signal $y(n)$ is also the output of the system.

Let us now find the autocorrelation function of the enhancer output signal $y(n)$. We make the standard assumption from independence theory which states that the filter weights and the input signal are independent [9].

The l-th autocorrelation lag of the filter output process $r(l) = E[y(n)y^*(n+l)]$, equals

$$r(l) = E[\mathbf{w}^H(n-1)\mathbf{x}(n-\Delta)\mathbf{x}^H(n-\Delta+l)\mathbf{w}(n-1+l)]. \tag{52}$$

The input signal $x(n)$ consists of two uncorrelated components $s(n)$, the sum of narrow band signals, and $v(n)$, the additive noise

$$x(n) = s(n) + v(n). \tag{53}$$

We can decompose the impulse response of the adaptive filter into two components. One of them is the optimal Wiener filter for the problem

$$\mathbf{w}_o = E[\mathbf{x}(n-\Delta)\mathbf{x}^H(n-\Delta)]^{-1} E[\mathbf{x}(n-\Delta)x(n)] \tag{54}$$

and the other one, $\tilde{\mathbf{w}}(n)$, represents the estimation errors.

$$\tilde{\mathbf{w}}(n) = \mathbf{w}_o - \mathbf{w}(n). \tag{55}$$

The output signal can hence be expressed as

$$y(n) = y_o(n) - \tilde{y}(n). \tag{56}$$

Substituting (53) and (55) into (52) and noticing that the cross-correlation between the Wiener filter output and that of the filter defined by weight errors is

$$E[y_o(n)\tilde{y}^*(l)] = E[\mathbf{w}_o^H \mathbf{x}(n-\Delta)\mathbf{x}^H(l-\Delta)\tilde{\mathbf{w}}(n-1)] = 0 \tag{57}$$

because of the adopted independence assumption and because $E[\tilde{\mathbf{w}}(n-1)] = E[\mathbf{w}_o - \mathbf{w}(n-1)] = 0$, we have

$$\begin{aligned}
r(l) = {} & E[\mathbf{w}_o^H \{\mathbf{s}(n-\Delta) + \mathbf{v}(n-\Delta)\}\{\mathbf{s}^H(n-\Delta+l) + \mathbf{v}^H(n-\Delta+l)\}\mathbf{w}_o] \\
& + E[\tilde{\mathbf{w}}^H(n-1)\{\mathbf{s}(n-\Delta) + \mathbf{v}(n-\Delta)\} \\
& \cdot \{\mathbf{s}^H(n-\Delta+l) + \mathbf{v}^H(n-\Delta+l)\}\tilde{\mathbf{w}}(n-1+l)].
\end{aligned} \tag{58}$$

Developing and grouping terms in the above equation results in

$$
\begin{aligned}
r(l) = \ & E[\mathbf{w}_o{}^H \mathbf{s}(n-\Delta)\mathbf{s}^H(n-\Delta+l)\mathbf{w}_o] \\
& + E[\mathbf{w}_o{}^H \mathbf{v}(n-\Delta)\mathbf{v}^H(n-\Delta+l)]\mathbf{w}_o] \\
& + E[\tilde{\mathbf{w}}^H(n-1)\mathbf{s}(n-\Delta)\mathbf{s}^H(n-\Delta+l)\tilde{\mathbf{w}}(n-1+l)] \\
& + E[\tilde{\mathbf{w}}^H(n-1)\mathbf{v}(n-\Delta)]\mathbf{v}^H(n-\Delta+l)\tilde{\mathbf{w}}(n-1+l)].
\end{aligned}
\tag{59}
$$

Using the fact that $\mathbf{w}_o$ is deterministic and the properties of the trace operator we further obtain

$$
\begin{aligned}
r(l) = \ & \mathbf{w}_o{}^H E[\mathbf{s}(n-\Delta)\mathbf{s}^H(n-\Delta+l)]\mathbf{w}_o \\
& + \mathbf{w}_o{}^H E[\mathbf{v}(n-\Delta)\mathbf{v}^H(n-\Delta+l)]\mathbf{w}_o \\
& + E[\mathrm{Tr}\{\tilde{\mathbf{w}}(n-1+l)\tilde{\mathbf{w}}^H(n-1)\mathbf{s}(n-\Delta)\mathbf{s}^H(n-\Delta+l)\}] \\
& + E[\mathrm{Tr}\{\tilde{\mathbf{w}}(n-1+l)\tilde{\mathbf{w}}^H(n-1)\mathbf{v}(n-\Delta)\mathbf{v}^H(n-\Delta+l)\}].
\end{aligned}
\tag{60}
$$

We now invoke the independence assumption saying that the weight vector $\tilde{\mathbf{w}}^H(n-1)$ is independent from the signals $\mathbf{s}(n-\Delta)$ and $\mathbf{v}(n-\Delta)$. This leads us to

$$
\begin{aligned}
r(l) = \ & \mathbf{w}_o{}^H E[\mathbf{s}(n-\Delta)\mathbf{s}^H(n-\Delta+l)]\mathbf{w}_o \\
& + \mathbf{w}_o{}^H E[\mathbf{v}(n-\Delta)\mathbf{v}^H(n-\Delta+l)]\mathbf{w}_o \\
& + \mathrm{Tr}\{E[\tilde{\mathbf{w}}(n-1+l)\tilde{\mathbf{w}}^H(n-1)]E[\mathbf{s}(n-\Delta)\mathbf{s}^H(n-\Delta+l)]\} \\
& + \mathrm{Tr}\{E[\tilde{\mathbf{w}}(n-1+l)\tilde{\mathbf{w}}^H(n-1)]E[\mathbf{v}(n-\Delta)\mathbf{v}^H(n-\Delta+l)]\}.
\end{aligned}
\tag{61}
$$

To proceed we need to find the matrix $\mathbf{K}(l) = E[\tilde{\mathbf{w}}(n-1+l)\tilde{\mathbf{w}}^H(n-1)]$.

5.1 Weight error correlation matrix

In this Section we investigate the combination of two adaptive filters and derive the expressions for the crosscorrelation matrix between the output signals of the individual filters $y_i(n)$ and $y_k(n)$. The autocorrelation matrices of the individual filter output signals follow directly using only one signal in the formulae.

For the problem at hands we can rewrite the equation (18) noting that we have introduced a Δ samples delay in the signal path as

$$
\tilde{\mathbf{w}}_i(n) \approx (\mathbf{I} - \mu_i \mathbf{R}_x)\tilde{\mathbf{w}}_i(n-1) - \mu_i \mathbf{x}(n-\Delta)e_o{}^*(n).
\tag{62}
$$

For the weight error correlation matrix we then have

$$\begin{aligned}
\mathbf{K}_{i,k,l}(n) = E[\tilde{\mathbf{w}}_i(n+l)\tilde{\mathbf{w}}_k^H(n)] &= E\big[(\mathbf{I}-\mu_i\mathbf{R_x})\tilde{\mathbf{w}}_i(n+l-1)\tilde{\mathbf{w}}_k(n-1)^H(\mathbf{I}-\mu_k\mathbf{R_x})\big] \\
&- E\big[(\mathbf{I}-\mu_i\mathbf{R_x})\tilde{\mathbf{w}}_i(n+l-1)\mu_k\mathbf{x}^H(n-\Delta)e_o(n)\big] \\
&- E\big[\mu_i\mathbf{x}(n-\Delta)e_o^*(n)(\mathbf{I}-\mu_k\mathbf{R_x})\tilde{\mathbf{w}}_k^H(n-1)\big] \\
&+ E\big[\mu_i\mu_k\mathbf{x}(n-\Delta)e_o^*(n)e_o(n)\mathbf{x}^H(n-\Delta)\big].
\end{aligned}$$

The second and third terms of the above equal zero because we have made the usual independence theory assumptions which state, that the weight errors $\tilde{\mathbf{w}}_i(n)$ are independent of the input signal $x(n\text{-}\Delta)$. To evaluate the last term we assume that the adaptive filters are long enough to remove all the correlation between $e_o(n)$ and $\mathbf{x(n\text{-}\Delta)}$. In this case we can rewrite the above as

$$\mathbf{K}_{i,k,l}(n) = (\mathbf{I}-\mu_i\mathbf{R_x})\mathbf{K}_{i,k,l}(n-1)(\mathbf{I}-\mu_k\mathbf{R_x}) + \mu_i\mu_k J_{min}\mathbf{R_x}, \tag{63}$$

where $J_{min} = E[|e_o|^2]$ is the minimum mean square error produced by the corresponding Wiener filter.

We now assume that the signal to noise ratio is low so that the input signal is dominated by the white noise process $v(n)$. In this case we can approximate the correlation matrix of the input process by unit matrix as

$$\mathbf{R_x} \approx \sigma_v^2\mathbf{I}, \tag{64}$$

where σ_v^2 is the noise variance. Later in the simulation study we will see that the theory developed this way actually works well with quite moderate signal to noise ratios. Then substituting (64) into (63) yields

$$\mathbf{K}_{i,k,l}(n) = (\mathbf{I}-\mu_i\sigma_v^2\mathbf{I})\mathbf{K}_{i,k,l}(n-1)(\mathbf{I}-\mu_k\sigma_v^2\mathbf{I}) + \mu_i\mu_k J_{min}\sigma_v^2\mathbf{I}. \tag{65}$$

In steady state, when $n \to \infty$ we have

$$\mathbf{K}_{i,k,l}(\infty) = (1-\mu_i\sigma_v^2)\mathbf{K}_{i,k,l}(\infty)(1-\mu_k\sigma_v^2) + \mu_i\mu_k J_{min}\sigma_v^2\mathbf{I}. \tag{66}$$

Solving the above for $\mathbf{K}_{i,k,l}(\infty)$ we have

$$\mathbf{K}_{i,k,l}(\infty) = \frac{\mu_i\mu_k J_{min}}{\mu_i\sigma_x^2 + \mu_k\sigma_x^2 - \mu_i\mu_k\sigma_v^2}\mathbf{I}. \tag{67}$$

5.2. Second order statistics of line enhancer output signal

As we see from the previous discussion, the correlation matrix of the weight error vector is diagonal. We therefore have that the matrix $\mathbf{K}_{i,k}(l) = E[\tilde{\mathbf{w}}_i(n-1+l)\tilde{\mathbf{w}}_k^H(n-1)]$ has in steady state, when $n \to \infty$, elements different form zero only alongside the main diagonal and the elements at this diagonal equal to $\frac{\mu_i \mu_k J_{min}}{\mu_i \sigma_x^2 + \mu_k \sigma_x^2 - \mu_i \mu_k \sigma_v^2}$. Substituting $\mathbf{K}_{i,k}(l)$ into (61) we now have that the l-th correlation lag of the output signal is equal to

$$
\begin{aligned}
r_{i,k}(l) = & \; \mathbf{w}_o^H E[\mathbf{s}(n-\Delta)\mathbf{s}^H(n-\Delta+l)]\mathbf{w}_o \\
& + \mathbf{w}_o^H E[\mathbf{v}(n-\Delta)\mathbf{v}^H(n-\Delta+l)]\mathbf{w}_o \\
& + r_s(l)N \frac{\mu_i \mu_k J_{min}}{\mu_i \sigma_x^2 + \mu_k \sigma_x^2 - \mu_i \mu_k \sigma_v^2} \\
& + \mathrm{Tr}\{\mathbf{K}_{i,k}(l)E[\mathbf{v}(n-\Delta)\mathbf{v}^H(n-\Delta+l)]\},
\end{aligned}
\tag{68}
$$

where $r_s(l)$ is the l-th autocorrelation lag of the input signal $s(n)$.

As the noise v has assumed to be white, the matrix $E[\mathbf{v}(n-\Delta)\mathbf{v}^H(n-\Delta+l)]$ has nonzero elements σ_v^2 only along the l-th diagonal and the rest of the matrix is filled with zeroes. Then

$$
\begin{aligned}
r_{i,k}(l) = & \; \mathbf{w}_o^H E[\mathbf{s}(n-\Delta)\mathbf{s}^H(n-\Delta+l)]\mathbf{w}_o \\
& + \sigma_v^2 \sum_{i=0}^{N-l-1} w_o^*(i)w_o(i+l) \\
& + r_s(l)N \frac{\mu_i \mu_k J_{min}}{\mu_i \sigma_x^2 + \mu_k \sigma_x^2 - \mu_i \mu_k \sigma_v^2} + r_0,
\end{aligned}
\tag{69}
$$

where $r_0 = N\sigma_v^2 \frac{\mu_i \mu_k J_{min}}{\mu_i \sigma_x^2 + \mu_k \sigma_x^2 - \mu_i \mu_k \sigma_v^2}$, if $l = 0$ and zero otherwise.

From (4) we see that the autocorrelation lags of the combination output signal $y(n)$ can be composed from its components $r_{i,k}(l)$ as follows

$$
\begin{aligned}
r(l) = & \; \lambda(n)^2 E[y_1(n)y_1^*(n+l)] + 2\lambda(n)(1-\lambda(n))E[Re\{y_1(n)y_2^*(n+l)\}] \\
& + (1-\lambda(n))^2 E[y_2(n)y_2^*(n+l)] \\
= & \; \lambda(n)^2 r_{1,1}(l) + 2\lambda(n)(1-\lambda(n))Re\{r_{1,2}(l)\} + (1-\lambda(n))^2 r_{2,2}(l).
\end{aligned}
$$

The autocorrelation matrix of y is a Toeplitz matrix $\mathbf{R}$ having the autocorrleation lags $r(l)$ along its first row.

Thus far we have evaluated the terms $E[y_i(n)y_k^*(n+l)]$, what remains is to find an expression for the steady state combination parameter $\lambda(\infty)$. For this purpose we can use (15), noting that $y_i(n) = \mathbf{w}_i^H(n-1)\mathbf{x}(n-\Delta)$. All the terms in the expression (15) are similar and we need to evaluate

$$\gamma_{ik} = E\left[\tilde{\mathbf{w}}_i^H(n-1)\mathbf{x}(n-\Delta)\mathbf{x}^H(n-\Delta)\tilde{\mathbf{w}}_k(n-1)\right]. \tag{70}$$

Due to the independence assumption we can rewrite (70) using the properties of trace operator as

$$\begin{aligned}
\gamma_{ik} &= Tr\left\{E\left[\mathbf{x}(n-\Delta)\mathbf{x}^H(n-\Delta)\tilde{\mathbf{w}}_k(n-1)\tilde{\mathbf{w}}_i^H(n-1)\right]\right\} \\
&= Tr\left\{\mathbf{R}_x E\left[\tilde{\mathbf{w}}_k(n-1)\tilde{\mathbf{w}}_i^H(n-1)\right]\right\} = Tr\left\{\mathbf{R}_x \mathbf{K}_{i,k,0}(n-1)\right\}.
\end{aligned} \tag{71}$$

We are now ready to find $\lambda(\infty)$ by substituting (71) and (67) into (15).

The power spectrum of the output process $y(n)$ is given by

$$P(f) = \lim_{K \to \infty} \frac{1}{K} E\,|\,[Y_K(f)|^2\,] = \sum_{l=-\infty}^{\infty} r(l)e^{-j2\pi l f}, \tag{72}$$

where $Y_K(f)$ is the length K discrete Fourier transform of the signal $y(n)$ and f is the frequency. There is a number of methods to compute an estimate of the power spectrum from the correlation matrix of a signal. In this paper we have used the Capon method [20].

$$\hat{P}(f) = \frac{K}{\mathbf{a}^H(f)\mathbf{R}^{-1}\mathbf{a}(f)}, \tag{73}$$

where $\mathbf{a}(f) = [1\,e^{-j2\pi f} \dots e^{-j2\pi(M-1)f}]^T$ and $\mathbf{R}$ is the $K \times K$ Toeplitz correlation matrix of the signal of interest. The Capon method was chosen because the signals we are interested in are sine waves in noise and the Capon method gives a more distinct spectrum estimate than the Fourier transform based methods in this situation.

6. Simulation Results

In this Section we present the results of our simulation study.

In order to obtain a practical algorithm, the expectation operators in both numerator and denominator of (7) have been replaced by exponential averaging of the type

$$P_{av}(n) = (1-\gamma)P_{av}(n-1) + \gamma p(n), \tag{74}$$

where $p(n)$ is the quantity to be averaged, $P_{av}(n)$ is the averaged quantity and γ is the smoothing parameter. The averaged quantities were then used in (7) to obtain λ. The curves shown in the Figures to follow are averages over 100 independent trials. We often show the simulation results and the theoretical curves in the same Figures. In several cases the curves overlap and are therefore indistinguishable.

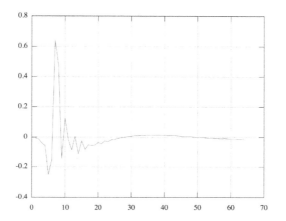

Figure 5. The true impulse response.

6.1 System Identification

We have selected the sample echo path model number one shown in Figure 5 from [10], to be the unknown system to identify and combined two 64 tap long adaptive filters.

In the Figures below the noisy blue line represents the simulation result and the smooth red line is the theoretical result. The curves are averaged over 100 independent trials.

In the system identification example we use Gaussian white noise with unity variance as the input signal. The measurement noise is another white Gaussian noise with variance $\sigma_v^2 = 10^{-3}$. The step sizes are $\mu_1 = 0.005$ for the fast adapting filter and $\mu_2 = 0.005$ for the slowly adapting filter. Figure 6 depicts the evolution of EMSE in time. One can see that the system converges fast in the beginning. The fast convergence is followed by a stabilization period between sample times 1000-7000 followed by another convergence to a lower EMSE level between the sample times 8000-12000. The second convergence occurs when the mean squared error of the filter with small step size surpasses the performance of the filter with large step size. One can observe that the there is a good accordance between the theoretical and the simulated curves so that the theoretical and the simulation curves are difficult to distinguish from each other.

The combination parameter λ is shown in Figure 7. At the beginning, when the fast converging filter gives smaller EMSE than the slowly converging one, λ is clise to unity. When the slow filter catches up the fast one λ starts to decrease and obtains a small negative value at the end of the simulation example. The theoretical and simulated curves fit well.

In the Figure 8 we show the time evolution of mean square deviation of the combination in the same test case. Again one can see that the theoretical and simulation curves fit well.

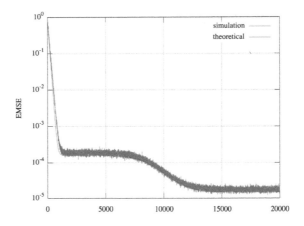

Figure 6. Time-evolutions of EMSE with $\mu_1 = 0.005$ and $\mu_2 = 0.0005$ and $\sigma_v^2 = 10^{-3}$.

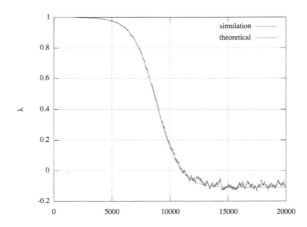

Figure 7. Time-evolutions of λ with $\mu_1 = 0.005$ and $\mu_2 = 0.0005$ and $\sigma_v^2 = 10^{-3}$.

6.2 Adaptive beamforming

In the beamforming example we have used a 8 element linear array with half wave-length spacing. The noise power is 10^{-4} in this simulation example. The useful signal which is 10 dB stronger than the noise arrives form the broadside of the array. There are three strong inter-ferers at $-35°$, $10°$ and $15°$ with $SNR_1 = 33$ dB and $SNR_2 = SNR_3 = 30$ dB respectively. The step sizes of the adaptive combination are $\mu_1 = 0.05$ and $\mu_2 = 0.006$.

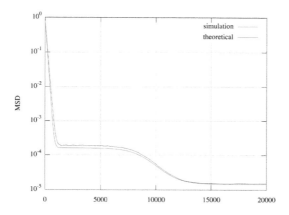

Figure 8. Time-evolutions of MSD with $\mu_1 = 0.005$ and $mu_{i2} = 0.0005$ and $\sigma_v^2 = 10^{-3}$.

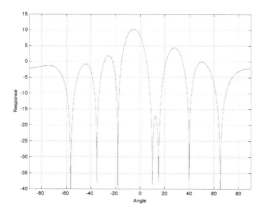

Figure 9. The antenna pattern.

The steady state antenna pattern is shown in Figure 6. One can see that the algorithm has formed deep nulls in the directions of the interferers while the response in the direction of the useful signal is equal to the number of antennas i.e. 8.

The evolution of EMSE in this simulation example is depicted in Figure 10. One can see a rapid convergence at the beginning of the simulation example. Then the EMSE value stabilizes at a certain level and after a while a second convergence occurs. The dashed red line is the theoretical result and the solid blue line is the simulation result. One can see that the two overlap and are indistinguishable in black and white print.

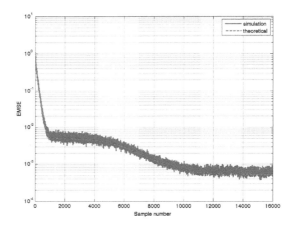

Figure 10. Time evolution of EMSE.

The time evolution of λ for this simulation example is shown in Figure 11. At the beginning λ is close to one forcing the output signal of the fast adapting filter to the output of the combination. Eventually the slow filter catches up with the fast one and λ starts to decrease obtaining at the end of the simulation example a small negative value so that the output signal is dominated by the output signal of the slowly converging filter. One can see that the simulation and theoretical curves for λ evolution are close to each other.

The signal to interference and noise ratio evolution is show in Figure 12. One can see a fast improvement of SINR at the beginning of the simulation example followed by a stabilization region. After a while a new region of SINR improvement occurs and finally the SINR stabilizes at an improved level. Again the theoretical result matches the simulation curve well making the curves indistinguishable in black and white print.

6.3. Adaptive Line Enhancer

In order to illustrate the adaptive line enhancer application we have used length $K = 32$ correlation sequences to form $K \times K$ correlation matrices for the Capon method. The narrow band signals were just sine waves in our simulations.

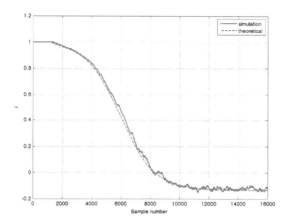

Figure 11. Time evolution of λ.

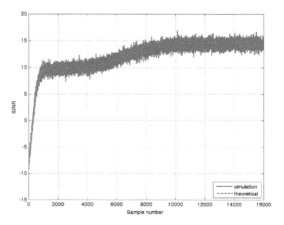

Figure 12. Time evolution of SINR.

The input signal consist of three sine waves and additive noise with unity variance. The sine waves with frequencies 0.1 and 0.4 have amplitudes equal to one and the third sine wave with normalized frequency 0.25 has amplitude equal to 0.5. The spectra of the input signal $x(n)$ and the output signal $y(n)$ are shown in Figure 13. The step sizes used were $\mu_1 = 0.5$ and $\mu_2 = 0.05$, the filter is $N=16$ taps long and the delay $\Delta = 10$.

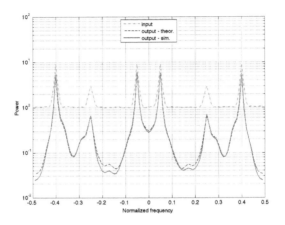

Figure 13. Line enhancer output signal spectrum.

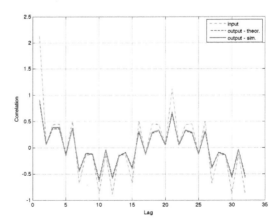

Figure 14. Line enhancer output signal autocorrelation.

In Figure 14 we show the correlation functions of input and output signals in the second simulation example. We can see that the theoretical correlation matches the correlation computed from simulations well.

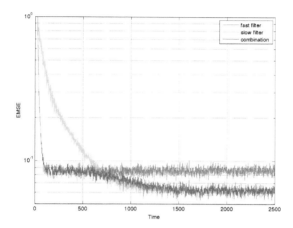

Figure 15. Evolution of EMSE of the two component filters and the combination in time.

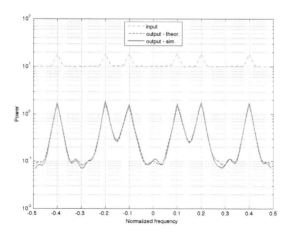

Figure 16. Line enhancer output signal spectrum.

The evolution of the excess mean square error of the combination together with that of the individual filters is shown in Figure 15. We see the fast initial convergence, which is due to the fast adapting filter. After the initial convergence there is a period of stabilization followed by a second convergence between the sample times 500 and 1500, when the error power of the slowly adapting filter bypasses that of the fast one.

In our final simulation example (Figure 16) we use three unity amplitude sinusoids with frequencies 0.1, 0.2 and 0.4. We have increased the noise variance to 10 so that the noise power is 20 times the power of each of the individual sinusoids. The adaptive filter is $N=16$ taps long and the delay $\Delta = 10$. The step sizes of the individual filters in the combined structure are $\mu_1 = 0.5$ and $\mu_2 = 0.005$. One can see that even in such noisy conditions there is still a reasonably good match between the theoretical and simulation results.

7. Conclusions

In order to make the LMS type adaptive algorithm work properly one has to select a suitable step size . The step size has to be smaller than $\frac{2}{\omega_{max}}$, where ω_{max} is the largest eigenvalue of the input signal autocorrelation matrix, in order to guarantee stability of the algorithm. Given that the stability condition is fulfilled a large step size allows the algorithm to initially converge fast but the mean square error in steady state remains large too. On the other hand if one selects a small step size it is possible to achieve a small steady state error but the initial convergence speed of the algorithm is reduced in this case. In this Chapter we have investigated the combination of two adaptive filters, which is a new and interesting way of achieving fast initial convergence and low steady state error of an adaptive filter at the same time, solving thus the trade-off one has in step size selection . We were looking at three applications of the technique - system identification, adaptive beamforming and adaptive line enhancing. In all three applications we saw that the combination worked as expected allowing the algorithm to converge fast to a certain level and then after a while providing a second convergence to a lower mean square error value.

Author details

Tõnu Trump[*]

Address all correspondence to: tonu.trump@gmail.com

Tallinn University of Technology, Estonia

References

[1] Aboulnasr, T., & Mayyas, K. (1997). A robust variable step-size LMS-type algorithm: Analysis and simulations. *IEEE Transactions on Signal Processing*, 45, 631-639.

[2] Arenas-Garcia, J., Figueiras-Vidal, A. R., & Sayed, A. H. (2006). Mean-square performance of convex combination of two adaptive filters. *IEEE Transactions on Signal Processing*, 54, 1078-1090.

[3] Armbruster, W. (1992). Wideband Acoustic Echo Canceller with Two Filter Structure, Signal Processing VI, Theories and Applications. VanderwalleJBoiteR.MoonenM.OosterlinckA., Elsevier Science Publishers B.V.

[4] Azpicueta-Ruiz, L. A., Figueiras-Vidal, A. R., & Arenas-Garcia, J. (2008a). A new least squares adaptation scheme for the affine combination of two adaptive filters. *Proc. IEEE International Workshop on Machine Learning for Signal Processing*, Cancun, Mexico, 327-332.

[5] Azpicueta-Ruiz, L. A., Figueiras-Vidal, A. R., & Arenas-Garcia, J. (2008b). A normalized adaptation scheme for the convex combination of two adaptive filters. *Proc. IEEE International Conference on Acoustics, Speech, and Signal Processing*, Las Vegas, Nevada, 3301-3304.

[6] Bershad, N. J., Bermudez, J. C., & Tourneret, J. H. (2008). An affine combination of two LMS adaptive filters - transient mean-square analysis. *IEEE Transactions on Signal Processing*, 56, 1853-1864.

[7] Fathiyan, A., & Eshghi, M. (2009). Combining several PBS-LMS filters as a general form of convex combination of two filters. *Journal of Applied Sciences*, 9, 759-764.

[8] Harris, R. W., Chabries, D. M., & Bishop, F. A. (1986). Variable step (vs) adaptive filter algorithm. *IEEE Transactions on Acoustics, Speech and Signal Processing*, 34, 309-316.

[9] Haykin, S. (2002). Adaptive Filter Theory, Fourth Edition,. Prentice Hall.

[10] ITU-T Recommendation G.168 Digital Network Echo Cancellers. (2009). *ITU-T*.

[11] Kim, K., Choi, Y., Kim, S., & Song, W. (2008). Convex combination of affine projection filters with individual regularization. *Proc. 23rd International Technical Conference on Circuits/Systems, Computers and Communications (ITC-CSCC)*, Shimonoseki, Japan, 901-904.

[12] Kwong, R. H., & Johnston, E. W. (1992). A variable step size LMS algorithm. *IEEE Transactions on Signal Processing*, 40, 1633-1642.

[13] Mandic, D., Vayanos, P., Boukis, C., Jelfs, B., Goh, S., I., Gautama, T., & Rutkowski, T. (2007). Collaborative adaptive learning using hybrid filters. *Proc. IEEE International Conference on Acoustics, Speech, and Signal Processing*, Honolulu, Hawaii, 901-924.

[14] Martinez-Ramon, M., Arenas-Garcia, J., Navia-Vazquez, A., & Figueiras-Vidal, A. R. (2002). An adaptive combination of adaptive filters for plant identification. *Proc. 14th International Conference on Digital Signal Processing*, Santorini, Greece, 1195-1198.

[15] Mathews, V. J., & Xie, Z. (1993). A stochastic gradient adaptive filter with gradient adaptive step size. *IEEE Transactions on Signal Processing*, 41, 2075-2087.

[16] Ochiai, K. (1977). Echo canceller with two echo path models. *IEEE Transactions on Communications*, 25, 589-594.

[17] Sayed, A. H. (2008). Adaptive Filters. John Wiley and sons.

[18] Shin, H. C., & Sayed, A. H. (2004). Variable step-size NLMS and affine projection algorithms. *IEEE Signal Processing Letters*, 11, 132-135.

[19] Silva, M. T. M., Nascimento, V. H., & Arenas-Garcia, J. (2010). A transient analysis for the convex combination of two adaptive filters with transfer of coefficients. *Proc. IEEE International Conference on Acoustics, Speech, and Signal Processing*, Dallas, TX, USA, 3842-3845.

[20] Stoica, P., & Moses, R. (2005). Spectral Analysis of Signals. Prentice Hall.

[21] Trump, T. (2009). An output signal based combination of two NLMS adaptive algorithms. *Proc. 16th International Conference on Digital Signal Processing*, Santorini, Greece.

[22] Trump, T. (2011a). analysisOutput signal based combination of two NLMS adaptive filters - transient, *Proceedings of the Estonian Academy of Sciences*, 60(4), 258-268.

[23] Trump, T. (2011b). Output statistics of a line enhancer based on a combination of two adaptive filters. *Central European Journal of Engineering*, 1, 244-252.

[24] Zhang, Y., & Chambers, J. A. (2006). Convex combination of adaptive filters for a variable tap-length LMS algorithm. *IEEE Signal Processing Letters*, 10, 628-631.

Chapter 3

On Using ADALINE Algorithm for Harmonic Estimation and Phase-Synchronization for the Grid-Connected Converters in Smart Grid Applications

Yang Han

Additional information is available at the end of the chapter

1. Introduction

The electric power transmission grid has been progressively developed for over a century, from initial design of local dc networks in low-voltage levels to three-phase high voltage ac networks, and finally to modern bulk interconnected networks with various voltage levels and multiple complex electrical components. The development of human society and economic needs is the major driving force the revolution of transmission grids stage-by-stage with the aid of innovative technologies. The current power industry is being modernized and tends to deal with the challenges more proactively by using the state-of-the-art technologies in the areas of sensing, communications, control, computing, and information technology. The shift in the development of transmission grids to be more intelligent has been summarized as "smart grid" [see Fig.1].

In a smart transmission network, flexible and reliable transmission capabilities can be facilitated by the advanced Flexible AC Transmission Systems (FACTS), high-voltage dc (HVDC) devices, and other power electronics-based devices. The FACTS devices are optimally placed in the transmission network to provide a flexible control of the transmission network and increase power transfer levels without new transmission lines. These devices also improve the dynamic performance and stability of the transmission network. Through the utilization of FACTS technologies, advanced power flow control, etc., the future smart transmission grids should be able to maximally relieve transmission congestions, and fully support deregulation and enable competitive power markets. In addition, with the increasing penetration of large-scale renewable/alternative energy resources, the future smart transmission grids would be able to enable full integration of these renewable energy re-sources(Wira et al., 2010, Sauter & Lobashov 2011, Varaiya et al., 2011).

Smart substations would provide advanced power electronics and control interfaces for re-newable energy and demand response resources so that they can be integrated into the pow-er grid on a large scale at the distribution level. By incorporating micro-grids, the substation can deliver quality power to customers in a manner that the power supply degrades grace-fully after a major commercial outage, as opposed to a catastrophic loss of power, allowing more of the installations to continue operations. Smart substations should have the capabili-ty to operate in the islanding mode taking into account the transmission capability, load de-mand, and stability limit, and provide mechanisms for seamlessly transitioning to islanding operation. Coordinated and self-healing are the two key characteristics of the next genera-tion control functions. These applications require precise tracking of the utility's phase-angle information, for high performance local or remote control, sensing and fault diagnosis pur-poses(Froehlich et al., 2011, Han et al., 2009).

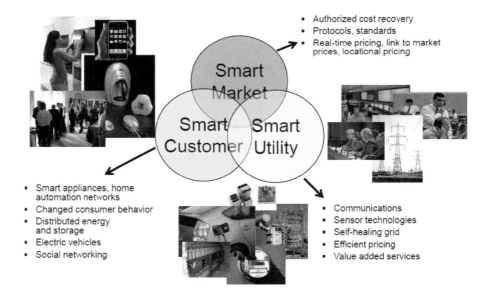

Figure 1. The vision of the future smart grid (SG) infrastructure

On the other hand, the proliferation of nonlinear loads causes significant power quality con-tamination for the electric distribution systems. For instance, high voltage direct transmis-sion (HVDC), electric arc furnaces (EAFs), variable speed ac drives which adopts six-pulse power converters as the first power conversion stage, these devices cause a large amount of characteristic harmonics and a low power factor, which deteriorate power quality of the electrical distribution systems. The increasing restrictive regulations on power quality problems have stimulated the fast development of power quality mitigation devices, which are connected to the grid to improve the energy transmission efficiency of the transmission lines and the quality of the voltage waveforms at the common coupling points (PCCs) for the customers. These devices are known as flexible AC transmission systems (FACTS) (Fig.2), which are based on the grid-connected converters and real-time digital signal processing techniques. Much work has been conducted in the past decades on the FACTS technologies and many FACTS devices have been practically implemented for the high voltage transmission grid, such as static synchronous compensators (STATCOMs), thyristor controlled series compensators (TCSCs) and unified power flow controllers (UPFCs) (Fig.3), etc(Cirrincione et al., 2008, Jarventausta et al, 2010).

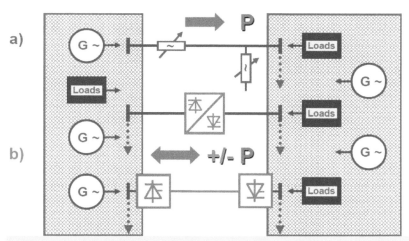

a) **FACTS: Voltage / Load-Flow Control (one Direction only) & POD**

b) HVDC Back-to-Back or Long-Distance Transmission:
   Voltage / Bidirectional Power-Flow Control, f-Control & POD

Figure 2. The circuit diagram of the FACTS and HVDC link

The stable and smooth operation of the FACTS equipments is highly dependent on how these power converters are synchronized with the grid. The need for improvements in the existing grid synchronization approaches also stems from rapid proliferation of distributed generation (DG) units in electric networks. A converter-interfaced DG unit, e.g., a photovoltaic (PV) unit (Fig.4), a wind generator unit (Fig.5) and a micro-turbine-generator unit, under both grid-connected and micro-grid (islanding) scenarios requires accurate converter synchronization under polluted and/or variable-frequency environment to guarantee stable operation of these grid-connected converters(Jarventausta et al., 2010).

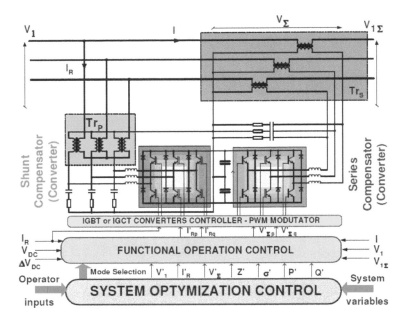

Figure 3. The circuit diagram of the unified power flow controller (UPFC)

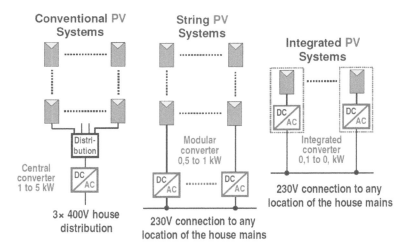

Figure 4. The configuration of PV arrays with the electric network

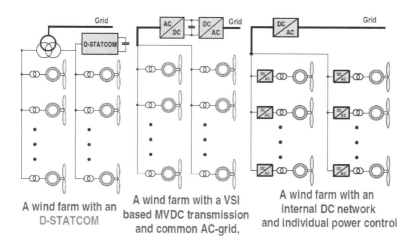

Figure 5. The configurations of the wind generators with the network

Besides, an active power filter (APF) (Fig.6) or dynamic voltage restorer (DVR) (Fig.7) rectifi-
er also requires a reference signal which is properly synchronized to the grid. Interfacing
power electronic converters to the utility grid, particularly at the medium and high voltages,
necessitates proper synchronization for the purpose of operation and control of the grid-
connected converters. However, the controller signals used for synchronization are often
corrupted by harmonics, voltage sags or swells, commutation notches, noise, phase-angle
jump and frequency deviations(Abdeslam et al., 2007, Cirrincione et al., 2008).

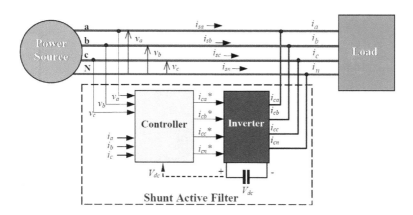

Figure 6. The circuit diagram of the shunt active power filter

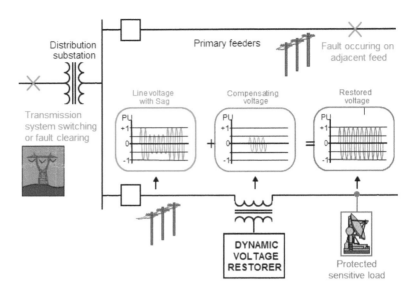

Figure 7. The circuit diagram of the dynamic voltage restorer (DVR)

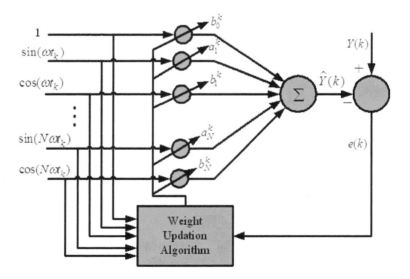

Figure 8. The diagram of the adaptive linear neural network (ADALINE)

Therefore, a desired synchronization method must detect the phase angle of the fundamental component of utility voltages as fast as possible while adequately eliminating the impacts of corrupting sources on the signal. Besides, the synchronization process should be updated not only at the signal zero-crossing, but continuously over the fundamental period of the signal(Chang et al., 2009, Chang et al., 2010). This chapter aims to present the harmonic estimation and grid-synchronization method using the adaptive linear neural network (ADALINE) (Figs.8 and 9). The mathematical derivation of these algorithms, the parameter design guidelines, and digital simulation results would be provided. Besides, their practical application for the grid-connected converters in smart grid would also be presented in this chapter.

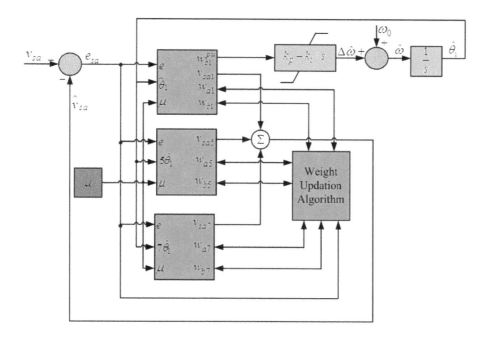

Figure 9. The grid-synchronization algorithm using the ADALINE-identifier

2. Mathematical model of the adaptive linear neural network (ADALINE)

The adaptive linear neural network (ADALINE) was used to estimate the time-varying magnitudes and phases of the fundamental and harmonics from a distorted waveform. The mathematical formulation of ADALINE is briefly reviewed. Consider an arbitrary signal $Y(t)$ with Fourier series expansion as (Simon, 2002):

$$Y(t) = \sum_{n=0,1,2,3,\cdots}^{N} A_n \sin(n\omega t + \varphi_n) + \varepsilon(t)$$

$$= \sum_{n=0,1,2,3,\cdots}^{N} (a_n \sin 2\pi n f t + b_n \cos 2\pi n f t) + \varepsilon(t),$$

$$(1)$$

where A_n and φ_n are correspondingly the amplitude and phase angle of the nth order harmonic component, and $\varepsilon(t)$ represents higher order components and random noise. In order to formulate the harmonic estimation problem by using ADALINE, we firstly define the pattern vector X_k and weight vector W_k as:

$$X_k = [1, \sin \omega t_k, \cos \omega t_k, \cdots, \sin N\omega t_k, \cos N\omega t_k]^T \tag{2}$$

$$W_k = [b_0^k, a_1^k, b_1^k, a_2^k, b_2^k, \ldots, a_N^k, b_N^k]^T \tag{3}$$

The square error on the pattern X_k is expressed as:

$$\varepsilon_k = \frac{1}{2}(d_k - X_k^T W_k)^2 = \frac{1}{2}e_k^2 = \frac{1}{2}(d_k^2 - 2d_k X_k^T W_k + W_k^T X_k X_k^T W_k) \tag{4}$$

where d_k is the desired scalar output. The mean-square error (MSE) ε can be obtained by calculating the expectation of both sides of Eq. (4), as:

$$\varepsilon = E[\varepsilon_k] = \frac{1}{2}E[d_k^2] - E[d_k X_k^T]W_k + \frac{1}{2}W_k^T E[X_k X_k^T]W_k \tag{5}$$

where the weights are assumed to be fixed at W_k while computing the expectation. The objective of the adaptive linear neural network (ADALINE) is to find the optimal weight vector W_k that minimizes the MSE of Eq. (4). For convenience of expression, Eq. (5) is rewritten as (Abdeslam et. al, 2007, Simon 2002):

$$\varepsilon = E[\varepsilon_k] = \frac{1}{2}E[d_k^2] - P^T W_k + \frac{1}{2}W_k^T R W_k \tag{6}$$

where P^T and R are defined as:

$$P^T = E[d_k X_k^T] = E[(d_k, d_k \sin \omega t_k, d_k \cos \omega t_k, \cdots, d_k \sin N\omega t_k, d_k \cos N\omega t_k)] \tag{7}$$

$$\mathbf{R} = E[X_k X_k^T] = E\begin{bmatrix} 1 & \sin\omega t_k & \cdots & \cos N\omega t_k \\ \sin\omega t_k & \sin\omega t_k \sin\omega t_k & \cdots & \sin\omega t_k \cos N\omega t_k \\ \cdots & \cdots & \cdots & \cdots \\ \cos N\omega t_k & \cos N\omega t_k \sin\omega t_k & \cdots & \cos N\omega t_k \cos N\omega t_k \end{bmatrix} \tag{8}$$

Notably, matrix R is real and symmetric, and ε is a quadratic function of weights. The gradient function $\nabla\varepsilon$ corresponding to the MSE function of Eq. (4) is obtained by straightforward differentiation:

$$\nabla\varepsilon = (\frac{\partial\varepsilon}{\partial b_0^k}, \frac{\partial\varepsilon}{\partial a_1^k}, \frac{\partial\varepsilon}{\partial b_1^k}, \cdots, \frac{\partial\varepsilon}{\partial a_N^k}, \frac{\partial\varepsilon}{\partial b_N^k})^T = -P + RW_k \tag{9}$$

which is a linear function of weights. The optimal set of weights, $\hat{W}_k$, can be obtained by setting $\nabla\varepsilon = 0$, which yields:

$$-P + R\hat{W}_k = 0 \tag{10}$$

The solution of the Eq. (10) is called Weiner solution or the Weiner filter:

$$\hat{W}_k = \mathbf{R}^{-1}P \tag{11}$$

The Weiner solution corresponds to the point in weight space that represents the minimum mean-square error ε_{min}. To compute the optimal filter one must first compute R⁻¹ and P. However, it would be difficult to compute R⁻¹ and P accurately when the input data comprises a random stream of patterns (drawn from a stationary distribution). Thus, by direct calculating gradients of the square error at the kth iteration:

$$\nabla\varepsilon_k = (\frac{\partial\varepsilon_k}{\partial b_0^k}, \frac{\partial\varepsilon_k}{\partial a_1^k}, \frac{\partial\varepsilon_k}{\partial b_1^k}, \cdots, \frac{\partial\varepsilon_k}{\partial a_N^k}, \frac{\partial\varepsilon_k}{\partial b_N^k})^T = e_k(\frac{\partial e_k}{\partial b_0^k}, \frac{\partial e_k}{\partial a_1^k}, \frac{\partial e_k}{\partial b_1^k}, \cdots, \frac{\partial e_k}{\partial a_N^k}, \frac{\partial e_k}{\partial b_N^k}) = -e_k X_k \tag{12}$$

where $e_k = (d_k - s_k)$, and $s_k = X_k^T W_k$ since we are dealing with linear neurons. Therefore, the recursive weights updating equation can be expressed as:

$$W_{k+1} = W_k + \mu(-\nabla\varepsilon_k) = W_k + \mu e_k X_k = W_k + \mu(d_k - s_k)X_k \tag{13}$$

where the learning rate μ is used to adjust the convergence speed and the stability of weights updating process. Taking the expectation of Eq. (12), the following equation is derived:

$$E[\widetilde{\nabla} \varepsilon_k] = -E[e_k X_k] = -E[d_k X_k - X_k X_k^T W_k] = RW_k - P = \nabla \varepsilon. \tag{14}$$

From Eq. (14), it can be found that the long-term average of $\widetilde{\nabla} \varepsilon_k$ approaches $\nabla \varepsilon$ hence $\widetilde{\nabla} \varepsilon_k$ can be used as unbiased estimate of $\nabla \varepsilon$. If the input data set is finite (deterministic), then the gradient $\nabla \varepsilon$ can be computed accurately by collecting the different $\widetilde{\nabla} \varepsilon_k$ gradients over all training patterns X_k for the same set of weights. The steepest descent search is guaranteed to search the Weiner solution provided the learning rate condition Eq. (15) is satisfied (Simon 2002):

$$0 < \mu < \frac{2}{\lambda_{max}} \tag{15}$$

where λ_{max} represents the largest eigenvalue of R. As for learning rate μ, increasing it results in a faster convergence at the trade-off of losing accuracy and increasing overshoots in transient response. Theoretically, a dynamical learning rate has better convergence characteristic, however, the implementation will be more demanding, and requires more expensive hardware setup. By a trial-and-error approach, a constant learning rate μ within the range of 0.025 and 0.04 is found sufficient for adequate stable convergence, which is consistent with Widrow-Hoff delta rule (Chang 2009, Chang 2010, Wira et al., 2010).

When mean-square error ε is minimized, the weight vector $\hat{W}$ after convergence would be:

$$\hat{W} = [b_0, a_1, b_1, a_2, b_2, ..., a_N, b_N]^T. \tag{16}$$

Thus the fundamental component of the measured signal $Y_1(t_k)$ is:

$$Y_1(t_k) = a_1 \sin \omega t_k + b_1 \cos \omega t_k. \tag{17}$$

Obviously, the dimension of the weight vector W_k to be updated depends on the order N of the harmonics to be estimated. In case of highly distorted load, lower order structure of neural network is not accurate enough when high convergence speed is required, so using higher order ANN structure is inevitable.

3. Synchronization for grid-connected converters using ADALINE technique

This Section formulates the generalized methodology for the phase-locked loop (PLL) design and synthesis by using adaptive linear neural network (ADALINE) technique. The mathematical derivation, the stability analysis and the detailed description of the proposed

ADALINE-PLL are outlined consecutively herein. In subsection 3.1, the optimal control parameters selection of the proposed ADALINE-PLL is discussed in terms of the continuous domain and the discrete domain analysis. Furthermore, the time-domain simulation results of the proposed ADALINE-PLL under different control parameters are also presented for verification.

3.1. Mathematical formulation of the ADALINE-PLL

This section presents the grid synchronization technique using the ADALINE algorithm. Firstly, the formulation of the ADALINE problem by using single-phase representation is outlined as follows. An arbitrary grid voltage can be represented as:

$$v_{sa}(t) = V_1 \sin(\omega_0 t + \varphi_1) + \sum_{n=2}^{N} V_n \sin(n\omega_0 t + \varphi_n) \tag{18}$$

where φ_1 and φ_n are the initial phase angle of the fundamental and nth order harmonic component, respectively. Here the dc offset is neglected for the sake of brevity. The phase angle of the fundamental component voltage can be expressed as:

$$\varphi_1 = \Delta\theta_1 + \theta_1 \tag{19}$$

where θ_1 and $\Delta\theta_1$ represent the estimated phase angle of the fundamental grid voltage and the estimation error, respectively, obtained from the ADALINE-PLL. Therefore, the phase angle of the nth order harmonic component can be expressed as:

$$n\omega_0 t + \varphi_n = n(\omega_0 t + \theta_1) + \varphi_n - n\theta_1 = n(\omega_0 t + \theta_1) + n\Delta\theta_1 + (\varphi_n - n\varphi_1) \tag{20}$$

where φ_n is the initial phase angle of the nth order harmonic component. Substituting Eq. (20) back into Eq. (19), rearranging terms, we get:

$$\begin{aligned} v_{sa}(t) &= V_1 \cos(\Delta\theta_1)\sin(\omega_0 t + \theta_1) + V_1 \sin(\Delta\theta_1)\cos(\omega_0 t + \theta_1) \\ &+ \sum_{n=2}^{N} \{V_n \cos(n\Delta\theta_1 + (\varphi_n - n\varphi_1))\sin[n(\omega_0 t + \theta_1)] \\ &\quad + V_n \sin(n\Delta\theta_1 + (\varphi_n - n\varphi_1))\cos[n(\omega_0 t + \theta_1)]\} \end{aligned} \tag{21}$$

From Eq. (21), it can be deduced that the original signal denoted by Eq. (18) can be regenerated by adjusting the coefficients $V_n \cos(n\Delta\theta_1 + (\varphi_n - n\varphi_1))$, $V_n \sin(n\Delta\theta_1 + (\varphi_n - n\varphi_1))$ (n=1, ..., N), even though the phase angle of the original signal is unknown. The objective of the proposed ADALINE-PLL is to reconstruct the phase information of the fundamental grid voltage φ_1 using least-mean-square (LMS) algorithm. Therefore, the grid voltage denoted by Eq.

(18) can be expressed by the inner product of two vectors, namely, the vector of trigonometric functions and the vector of weights in the LMS-based weights updating algorithm. The weight vector W is denoted by the coefficients of the corresponding trigonometric functions. Followed by this idea, Eq. (21) can be expressed as:

$$\hat{Y} = W^T X \tag{22}$$

where $\hat{Y}$ is the estimated output of the grid voltage $v_{sa}(t)$ by using the LMS-based linear optimal filter methodology. The vector W and X corresponding to the weight vector and the input vector, respectively, are represented as:

$$W = [V_1 \cos(\Delta\theta_1), V_1 \sin(\Delta\theta_1), ..., V_n \cos(n\Delta\theta_1 + (\varphi_n - n\varphi_1)), V_n \sin(n\Delta\theta_1 + (\varphi_n - n\varphi_1))]^T \tag{23}$$

$$X = [\sin(\omega_0 t + \theta_1), \cos(\omega_0 t + \theta_1), ..., \sin[n(\omega_0 t + \theta_1)], \cos[n(\omega_0 t + \theta_1)]]^T \tag{24}$$

Equation (23) can be rewritten as:

$$W = [\omega_{a1}, \omega_{b1}, ..., \omega_{aN}, \omega_{bN}]^T \tag{25}$$

Notably, the salient difference between the ADALINE algorithm and the ADALINE-PLL algorithm is that, the frequency and phase angle signals utilized in the ADALINE weights updating process were assumed to be constant. However, in case of the ADA-LINE-PLL, the frequency and phase angle of fundamental component grid voltage is recursively updated by the loop filter (LF) and voltage controlled oscillator (VCO) of the PLL. In other words, the weights updating procedure of the ADALINE is utilized as the phase detector (PD) for the PLL, which generate the error signal to drive the loop filter (LF) and voltage controlled oscillator (VCO), according to the initial definition of PLL The graphical interpretation of the proposed ADALINE-PLL is illustrated in Fig.9. In order to better illustrate the working principle of the proposed ADALINE-PLL, the weights updating law and stability conditions are discussed in detail as follows.

In the discrete domain, the weight vector of the ADALINE should be changed in a minimum manner, subject to the constraint imposed on the updated filter output. Let $\hat{W}_k$ denote the old weight vector of the ADALINE filter at the kth iteration and $\hat{W}_{k+1}$ denote its updated weight vector at the $(k+1)$th iteration. Therefore, given the input vector X_k and the desired output Y_k, the weight vector $\hat{W}_{k+1}$ can be written as:

$$\delta \hat{W}_{k+1} = \hat{W}_{k+1} - \hat{W}_k \tag{26}$$

For each (X_k, Y_k) pair, there exist at least one $\hat{W}_{k+1}$, such that the following equation is satisfied:

$$\hat{W}_{k+1}^H X_k = Y_k \tag{27}$$

Hence the weights adaption process is achieved by solving the optimization problem, as indicated by Eqs. (26)-(27). The cost function at the kth iteration can be formulated by using the method of Lagrange multipliers (Wira et al., 2010, Yin et al., 2010), as:

$$J_k = || \delta \hat{W}_{k+1} ||^2 + \lambda \cdot (Y_k - \hat{W}_{k+1}^H X_k) \tag{28}$$

where λ denotes the real-valued Lagrange multiplier. The term $|| \delta \hat{W}_{k+1} ||^2$ denotes the squared Euclidean norm of the weight change δW_{k+1}. The cost function is a quadratic function of the weight vector $\hat{W}_{k+1}$, as shown by expanding Eq. (28) into:

$$J_k = (\hat{W}_{k+1} - \hat{W}_k)^H \cdot (\hat{W}_{k+1} - \hat{W}_k) + \lambda \cdot (Y_k - \hat{W}_{k+1}^H X_k) \tag{29}$$

The optimum weight vector can be found by minimizing the cost function J_k. Differentiate the cost function J_k with respect to $\hat{W}_{k+1}$, we get:

$$\frac{\partial J_k}{\partial \hat{W}_{k+1}} = 2(\hat{W}_{k+1} - \hat{W}_k) - \lambda X_k \tag{30}$$

By setting Eq.(30) equal to zero, the optimum value for $\hat{W}_{k+1}$, corresponding to the stationary point of the cost function J_k, can be derived as:

$$\hat{W}_{k+1} = \hat{W}_k + \frac{1}{2} \lambda X_k \tag{31}$$

Hence, the output of the ADALINE as denoted by Eq. (22) can be rewritten as:

$$Y_k = \hat{W}_{k+1}^H X_k = (\hat{W}_k + \frac{1}{2} \lambda X_k)^H X_k = \hat{W}_k^H X_k + \frac{1}{2} \lambda || X_k ||^2 \tag{32}$$

Then, the Lagrange multiplier λ can be obtained as:

$$\lambda = \frac{2e_k}{||X_k||^2} \tag{33}$$

where $e_k = Y_k - \hat{W}^H_k X_k$ represents the estimation error of the ADALINE. From Eq. (31) and Eq. (32), the following equation can be derived:

$$\delta\hat{W}_{k+1} = \hat{W}_{k+1} - \hat{W}_k = \frac{1}{||X_k||^2}X_k e_k \tag{34}$$

In order to ensure stable operation of the weight vector updating process, a positive real scaling factor μ (learning rate) is introduced to the step size. Hence Eq. (34) can be redefined as:

$$\delta\hat{W}_{k+1} = \hat{W}_{k+1} - \hat{W}_k = \frac{\mu}{||X_k||^2}X_k e_k \tag{35}$$

Equivalently,

$$\hat{W}_{k+1} = \hat{W}_k + \frac{\mu}{||X_k||^2}X_k e_k \tag{36}$$

The aforementioned weights updating scheme, in essence, belongs to the well-known least mean square (LMS) algorithm, which may introduce convergence problem in case of small input vector X_k since the squared norm $||X_k||^2$ appears in the denominator, as indicated by Eq. (36). To solve this problem, Eq. (36) can be modified as (Chang 2009):

$$\hat{W}_{k+1} = \hat{W}_k + \frac{\mu}{\delta+||X_k||^2}X_k e_k \tag{37}$$

where δ is a sufficiently small real number and $\delta > 0$. The weight adaptation law represented in Eq.(37) is adopted and practically implemented herein.

3.2. Stability analysis of the ADALINE

The selection of the step-size parameter μ is a compromise between the estimation accuracy and the convergence speed of the weights updating process. Generally speaking, a higher step-size would result in faster dynamic response and wider bandwidth of the ADALINE-PLL. On the other hand, if the step-size is selected too small, the corresponding ADALINE would be slow in transient response and results in a narrow bandwidth in frequency domain. Assuming that the physical mechanism responsible for generating the desired response Y_k is controlled by the multiple regression model:

$$Y_k = \hat{W}_{k+1}^H X_k = W^H X_k + d_k \tag{38}$$

where W represents the model's unknown parameter vector and d_k represents unknown disturbances that accounts for various system impairments, such as random noise, modeling errors or other unknown sources. The weight vector $\hat{W}_k$ computed by the ADALINE algorithm is an estimate of the actual weight vector W, hence the estimation error can be presented by:

$$\varepsilon_k = W - \hat{W}_k \tag{39}$$

From Eqs.(37)-(39), the incremental in the estimation error can be derived as:

$$\varepsilon_{k+1} = \varepsilon_k - \frac{\mu}{\delta + ||X_k||^2} X_k e_k \tag{40}$$

As stated above, the underlying idea of the ADALINE design is to minimize the incremental change in the weight vector $\hat{W}_{k+1}$ from the kth and $(k+1)$th iteration, subject to a constraint imposed on the updated weight vector $\hat{W}_{k+1}$. Based on this idea, the stability of the ADALINE algorithm can be investigated by defining the mean-square deviation of the weight vector estimation error, hence we get:

$$\rho_n = E[||\varepsilon_k||^2] \tag{41}$$

Taking the squared Euclidean norms of both sides of Eq. (41), rearranging terms, and then taking the expectations on both sides of equation, we get:

$$\rho_{n+1} = \rho_n + \mu^2 E[\frac{||X_k||^2 \cdot |e_k|^2}{(\delta + ||X_k||^2)^2}] - 2\mu E[\frac{\xi_k e_k}{\delta + ||X_k||^2}] \tag{42}$$

where ξ_k denotes the undisturbed error signal defined by

$$\xi_k = (W - \hat{W}_k)^H X_k = \varepsilon_k^H X_k \tag{43}$$

From Eq.(43), it shows that the mean-square deviation ρ_n decrease exponentially with the increase of iterations, hence the ADALINE is therefore stable in the mean-square error sense (i.e., the convergence process is monotonic), provided $\rho_{n+1} < \rho_n$ is satisfied, which corresponding to the following condition:

$$0 < \mu < 2 \frac{E[\xi_k e_k / (\delta + ||X_k||^2)]}{E\{||X_k||^2 \cdot |e_k|^2 / [(\delta + ||X_k||^2)^2]\}} \tag{44}$$

Considering the limited rate of variation in parameters for the practical grid-connected converter applications, if faster adaptation for the weight vector $\hat{W}_{k+1}$ than the parameter variation of the input signal is ensured, it can be shown that this inequality can always be satisfied. It should be noted that the selection of the step-size parameter μ has a significant effect on the frequency characteristics of the ADALINE-PLL, which would be discussed in the forthcoming subsection. Here we first describe the proposed ADALINE-PLL and its implementation in Matlab/Simulink[1].

3.3. Description of the proposed ADALINE-PLL

Figs.10-11 show the single-phase and three-phase version of the proposed ADALINE-PLL. The following discussion is mainly focused on the single-phase version of the ADALINE-PLL, but the similar analysis can be easily extended to the three-phase version. For the sake of brevity, only the fundamental component, fifth and seventh order harmonics are considered in the grid voltages, hence the estimation blocks corresponding to these three components are considered in the single-phase ADALINE-PLL. One may extend the order of the ADALINE-PLL by incorporating higher order harmonic blocks in the algorithm according to the particular applications. Fig.10(a) shows the top layer representation of the single-phase ADALINE-PLL, it can be observed that the estimation error, phase angle of the fundamental component in grid voltage, the learning rate are utilized as the input signals to the subsystems, namely, the fundamental frequency block, the fifth order harmonic block and the seventh order harmonic block.

Figs.10(b)-(d) shows the three subsystems for individual harmonic component estimation, namely, the fundamental component, the fifth and the seventh order harmonic components. Once again, the weights of the fundamental frequency component are denoted as ω_{a1} and ω_{b1}, hence the phase estimation error denoted by $\Delta\theta_1$ can be regulated to zero by using a properly designed closed-loop control system, which resembles that of the existing grid synchronization schemes. As shown in Fig.10(b), the per unit representation of the weight ω_{b1} is

---

1 www.mathworks.com

utilized as the input signal for the loop filter (LF) of the PLL, which can be simply derived
as:

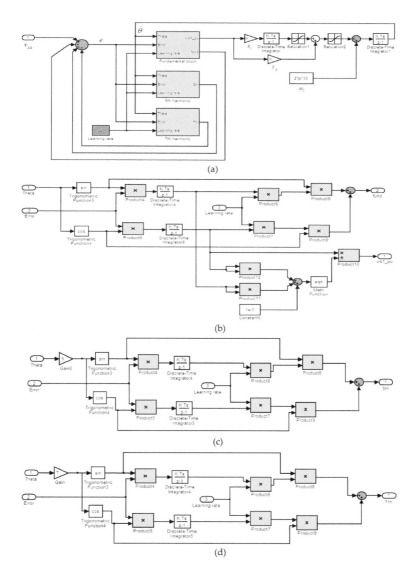

(a)

(b)

(c)

(d)

Figure 10. The Matlab/Simulink diagram for the single-phase ADALINE-PLL

$$\omega_{b1}^{pu} = \frac{\omega_{b1}}{\sqrt{\omega_{a1}^2 + \omega_{b1}^2}} = \frac{V_1 \sin(\Delta\theta_1)}{\sqrt{V_1^2 \sin^2(\Delta\theta_1) + V_1^2 \cos^2(\Delta\theta_1)}} = \sin(\Delta\theta_1) \tag{45}$$

The derived signal ω_{b1}^{pu}, is then used as input for the phase tracking algorithm. However, by incorporating the adaptive linear optimal filter methodology, the proposed ADALINE-PLL exhibits noticeable advantages compared to the existing grid synchronization algorithms in terms of response speed, accuracy and robustness.

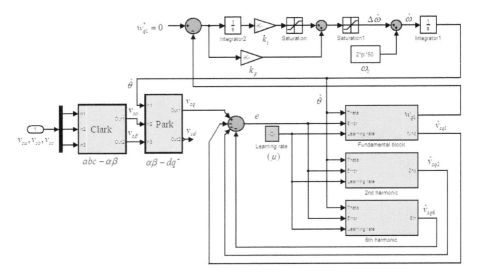

Figure 11. The Matlab/Simulink diagram for the three-phase ADALINE-PLL

Fig. 11 shows the corresponding three-phase version of the proposed ADALINE-PLL, which has a similar architecture with that of the single-phase version. One of the salient features of the three-phase ADALINE-PLL algorithm is that the Clark's transformation and Park's transformation are utilized consecutively to derive the q-axis component of the grid voltages, similar to the procedure adopted in the conventional three-phase PLL (CPLL) and the virtual PLL (VPLL). However, the adaptive linear optimal filter (ADALINE) is used as the phase detector (PD) section, which generate the dc component for the voltage controlled oscillator (PI regulator). It should be noted that there is one fundamental frequency shift when the electric quantities are transformed from the stationary α-β reference frame to the synchronous rotating reference frame (d-q frame). Besides, it is well known that the typical balanced nonlinear load produce characteristic harmonics of the orders: -5, +7, -11, +13... $6n+1$ (n is integer), corresponding to the $6n$th order harmonic components in synchronous rotating reference frame. Therefore, the 2nd order harmonic in Fig.11 corresponds to the fundamental frequency negative sequence component, while the 6th order harmonic corresponds to the 5th order harmonic (negative sequence) and the 7th order harmonic (positive sequence) in stationary phase a-b-c frame. Generally speaking, the harmonic components considered in the proposed ADALINE-PLL are selected according to the particular applications and the available computational resources.

3.4. Parameter selection of the ADALINE-PLL

In this section, the parameter design of the single-phase version ADALINE-PLL is discussed by using continuous domain (s-domain) analysis, discrete domain (z-domain) analysis and time-domain simulation. It is found that the proposed ADALINE-PLL has the characteristic of band-pass filter around the fundamental frequency and a notch filter at harmonic frequencies.

3.4.1. Continuous-domain (s-domain) analysis

Assuming the phase angle of the fundamental grid voltage detected by the closed-loop ADALINE-PLL is denoted by $\hat{\theta}$, which is an integral of the estimated angular frequency $\hat{\omega}_0$. In the steady state, the estimated angular frequency $\hat{\omega}_0$ can be considered to be constant, hence the phase angle can be approximated as $\hat{\theta} = \hat{\omega}_0 t$. Therefore, the block diagram of the ADALINE-PLL indicated by Fig.10 can be simplified as Fig.12, provided that the estimated angular frequency $\hat{\omega}_0$ is within its neighborhood, i.e., $\hat{\omega}_0' \leq \hat{\omega}_0 \leq \hat{\omega}_0''$ ($\hat{\omega}_0'$ and $\hat{\omega}_0''$ represent the lower and upper boundaries which defines the lock range of the PLL). Referring to the fundamental frequency block in Fig.12, the estimated fundamental component in time domain can be represented as:

$$v_{sa1}(t) = \left\{ [e(t) \cdot \cos(\hat{\omega}_0 t)] * h_1(t) \right\} \cdot \cos(\hat{\omega}_0 t) + \left\{ [e(t) \cdot \sin(\hat{\omega}_0 t)] * h_1(t) \right\} \cdot \sin(\hat{\omega}_0 t) \qquad (46)$$

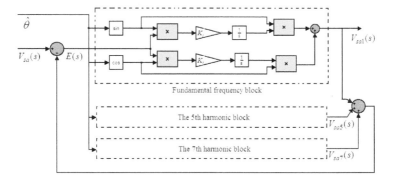

Figure 12. Frequency domain diagram for quasi-steady state analysis of the ADALINE-PLL

where $e(t)$ represents the estimation error of the ADALINE, $v_{sa1}(t)$ represents the estimated fundamental component of grid voltage, $h_1(t)$ represents the operator of integration and asterisk denotes convolution. Applying Laplace transform to Eq. (46), rearranging terms, we get:

$$V_{sa1}(s) = \frac{1}{2}[H_1(s + j\hat{\omega}_0) + H_1(s - j\hat{\omega}_0)] \cdot E(s) \tag{47}$$

where $V_{sa1}(s)$, $H_1(s)$, $E(s)$ corresponds to the Laplace transform of $v_{sa1}(t)$, $h_1(t)$ and $e(t)$, respectively. In Eq. (47), $H_1(s)$ is represented as:

$$H_1(s) = \frac{k_1}{s} \tag{48}$$

where k_1 is integration gain, corresponding to the learning rate (μ) of the weights updating process ($\mu = k_1 T$). Combining Eq.(47) and Eq.(48), we get

$$G_1(s) = \frac{V_{sa1}(s)}{E(s)} = \frac{k_1 s}{s^2 + \hat{\omega}_0^2} \tag{49}$$

Similarly, for the nth order harmonic block in Fig.12, the generalized transfer function from estimation error $E(s)$ to the individual harmonic component output $V_{san}(s)$, can be derived as:

$$G_n(s) = \frac{V_{san}(s)}{E(s)} = \frac{k_n s}{s^2 + (n\hat{\omega}_0)^2} \tag{50}$$

For the present case, the fundamental component, fifth and seventh order harmonics are considered, hence the error transfer function from the input $V_{sa}(s)$ to $E(s)$, can be represented as:

$$G_{error}(s) = \frac{E(s)}{V_{sa}(s)} = \frac{1}{1 + G_1(s) + G_5(s) + G_7(s)} \tag{51}$$

Similarly, the transfer function from the input $V_{sa}(s)$ to the estimated fundamental component $V_{sa1}(s)$, is:

$$G_{fund}(s) = \frac{V_{sa1}(s)}{V_{sa}(s)} = \frac{G_1(s)}{1 + G_1(s) + G_5(s) + G_7(s)} \tag{52}$$

Fig. 13 shows the bode-plot of the ADALINE when only the fundamental frequency block is considered. The frequency response of the ADALINE under the variations of the center frequency $\hat{\omega}_0$ and the integration gain are shown in Fig.13(a) and Fig.13(b), respectively. Fig. 13(a) shows the open-loop frequency response of the ADALINE with the variation of center frequency, it is interesting to notice that this characteristic provides the flexible frequency tracking capability, compared to the adaptive linear neural network (ADALINE) algorithm since the frequency response of ADALINE cannot adapt to the frequency variation in the input signal. It can be observed from Fig.13(b) that the integration gain, i.e., the learning rate (μ), has a significant effect on the frequency characteristics of the ADALINE. Small learning rate results in a sharp amplitude-frequency curve and steep phase-frequency curve. Besides, small learning rate implies a narrow bandwidth and slow transient response of the weights updating process. Higher learning rate, on the other hand, implies a flat amplitude-frequency curve, which would improve the dynamic response, increase the bandwidth of the ADALINE.

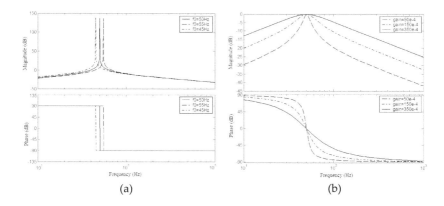

(a)   (b)

Figure 13. Bode plot of the ADALINE when only the fundamental frequency block is considered. (a) Open-loop frequency response of ADALINE with the variation of the center frequency, (b) Closed-loop frequency response of ADALINE with the variation of gain.

Fig.14 shows the frequency response of the ADALINE when the fundamental component, fifth and seventh harmonic components are considered. Fig.14(a) shows the bode-plot from the input signal $V_{sa}(s)$ to the estimation error $E(s)$. It can be observed that it exhibits as a typical notch filter, and significant attenuation is observed in the amplitude-frequency curve at the harmonic components under consideration. The attenuation at particular harmonic frequency is controlled by the selection of the learning rate of ADALINE, higher learning rate implies higher attenuation. Fig.14(b) shows the bode-plot from the input signal $V_{sa}(s)$ to the estimated fundamental component $V_{sa1}(s)$. It can be observed that it exhibits a band-pass filter around the fundamental frequency, and a notch filter at the considered harmonic frequencies. In case of large frequency variation in grid voltages, the learning rates of the ADALINE should be sufficiently high to ensure a wide bandwidth. Besides, it should be noted that the number of harmonics considered in the ADALINE-PLL can be easily extended to higher order harmonic components according to the particular applications.

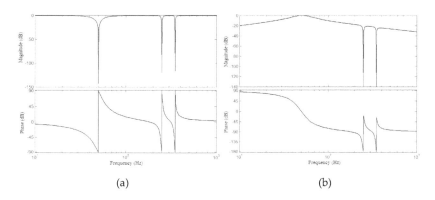

(a)   (b)

Figure 14. Bode plot of the ADALINE when the fundamental frequency block, the fifth and seventh harmonic blocks are considered.

It should be noted that the frequency domain analysis is based on the quasi-steady state model of the ADALINE, which serves the purpose of phase detection (PD) for the PLL. The estimated phase error signal is then utilized as the input for the loop filter (LF), which is selected as the standard proportional-integral (PI) regulator for the present case. Here the linearized model for the phase estimation can be described as Fig.15(a). It is interesting to observe that the derived linearized model for the phase estimation resembles that of the existing PLL algorithms. The closed-loop transfer function of the linearized model indicated by Fig.15(a) can be represented as:

$$H_c(s) = \frac{\hat{\theta}(s)}{\theta(s)} = \frac{K_f(s)}{s + K_f(s)} \tag{53}$$

where $\hat{\theta}(s)$, $\theta(s)$ denote the Laplace transform of the estimated phase angle $\hat{\theta}$ and the actual phase angle θ respectively. To achieve a good trade-off between the filter performance and system stability, the proportional-integral (PI) type filter is utilized for the loop filter (LF), which can be given as:

$$K_f(s) = k_p(1 + \frac{1}{\tau s}) \tag{54}$$

where k_p and τ denote the proportional gain and time constant of the PI regulator, and the integrator gain $k_i = k_p/\tau$. Equation (54) can be rewritten in the generalized second order system as:

$$H_c(s) = \frac{2\xi\omega_n s + \omega_n^2}{s^2 + 2\xi\omega_n s + \omega_n^2} \tag{55}$$

where

$$\omega_n = \sqrt{k_p/\tau}, \xi = \frac{k_p}{2\omega_n} = \frac{\sqrt{\tau k_p}}{2} \tag{56}$$

The open loop transfer function of Fig.15 (a) can be derived as:

$$G_{open} = \frac{K_f(s)}{s} = \frac{k_p(1 + \frac{1}{\tau s})}{s} = \frac{k_p(s + \frac{1}{\tau})}{s^2} \tag{57}$$

The root locus for the PLL modeled in the s-domain is shown in Fig.15(b). There are two open loop poles at the origin of the s-plane and one open loop zero at s=-1/τ. However, it is interesting to notice from Fig.15(b) that the s-domain model never predicts an unstable mode for any combination of PI parameters. Therefore, the discrete domain (z-domain) would be necessary to study the stability characteristic of the proposed ADALINE-PLL, as discussed in subsequent section.

3.4.2. Discrete-domain (z-domain) analysis

In the discrete domain, Eq. (50) can be rewritten as:

$$G_n(z) = \frac{V_{san}(z)}{E(z)} = \frac{k_n z(z - \cos\Omega_n)}{z^2 - 2z\cos\Omega_n + 1} \tag{58}$$

where $\Omega_n = n\hat{\omega}_0 T$, and T is the sampling frequency specified according to the particular applications, for the present case, $T=100\mu s$ is selected which is the typical sampling frequency for the low voltage power converters. Hence, the discrete domain transfer function from $V_{sa}(z)$ to $E(z)$ can be represented as:

$$G_{error}(z) = \frac{E(z)}{V_{sa}(z)} = \frac{1}{1+G_1(z)+G_5(z)+G_7(z)} = \frac{1}{1 + \displaystyle\sum_{n=1,5,7} \frac{k_n z(z - \cos\Omega_n)}{z^2 - 2z\cos\Omega_n + 1}} \qquad (59)$$

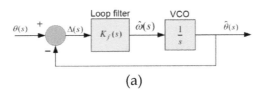

(a)

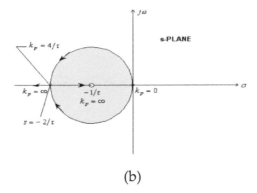

(b)

Figure 15. Small signal analysis of the proposed ADALINE-PLL in s-domain: (a) The approximated second order linearized model for phase estimation and (b) Root locus in s-domain of the linearized model.

Assuming that $G(z)=G_1(z)+G_5(z)+G_7(z)$, $\hat\omega_0=2\times\pi\times50$, $T=100\mu s$, and the integration gain k_n of individual harmonic component are assumed to be identical for the sake of simplicity ($k_n=K$), then the following representation can be derived:

$$G(z) = K\frac{0.0002988z^5 - 0.001479z^4 + 0.002944z^3 - 0.002944z^2 + 0.001479z - 0.0002988}{z^6 - 5.926z^5 + 14.71z^4 - 19.56z^3 + 14.71z^2 - 5.926z + 1} \qquad (60)$$

The root locus for the ADALINE modeled in the z-domain is shown in Fig.16. There are two open loop zeros at z=1, a pair of conjugate zeros and three pair of conjugate poles distributing in the z-plane. It can be observed from Fig.16 that the stability margin increases with the increase of integration gain K when 80<K<554 (0.008<μ<0.055) and decreases with the increase of K when 554<K<6833 (0.055<μ<0.68). Moreover, it can be observed from the root locus diagram that when 80<K<6833 (0.008<μ<0.68), the ADALINE system is stable, otherwise it is unstable.

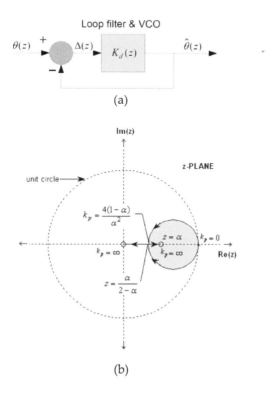

(a)

(b)

Figure 16. Small signal analysis of the proposed ADALINE-PLL in z-domain: (a) The approximated second order linearized model for phase estimation and (b) Root locus in z-domain of the linearized model.

The ADALINE subsystem is assumed to be stable in the following discrete domain analysis, which implies that the phase detection is achieved. The z-domain analysis will be performed on a discrete-time PLL system with a second-order loop filter. As shown in Fig.17(a), and the block $K_d(z)$ is the z-transform of the loop filter and voltage-controlled oscillator (VCO), hence the closed-loop transfer function can be represented as:

$$H_c(z) = \frac{\hat{\theta}(z)}{\theta(z)} = \frac{K_d(z)}{1 + K_d(z)} \tag{61}$$

For the second order loop using the PI type filter, $K_d(z)$ can be obtained as

$$K_d(z) = k_p \frac{z(z-\alpha)}{(z-1)^2} \tag{62}$$

where $\alpha = 1 - T/\tau$ and T denotes the sampling period of the discrete system. The transfer function of the closed loop system in the discrete-time domain can be derived by substituting Eq. (61) into Eq. (60) as

$$H_c(z) = H_{cm} \frac{z(z-\alpha)}{z^2 - az + b} \tag{63}$$

where

$$H_{cm} = \frac{k_p}{1+k_p}, a = \frac{2+k_p\alpha}{1+k_p}, b = \frac{1}{1+k_p} \tag{64}$$

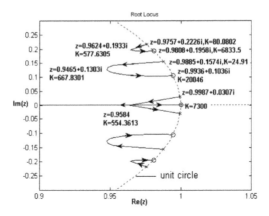

Figure 17. Root locus of discrete-time ADALINE system

The root locus for the PLL modeled in the z-domain is shown in Fig.17(b). It can be observed that there are two open loop poles at $z=1$ and two open loop zeros at $z=0$ and $z=\alpha$. It is interesting to note that, since the open-loop zero location (α) is a function of the time constant τ, the z-domain model can predict unstable loop performance for the condition of $T > 2\tau$ in which case an open-loop zero α is located on the negative real axis outside the unit circle. For $T << \tau$, the quantity α is close to unity, in this case, the z-domain and s-domain model predict similar characteristics for jitter[2] frequencies within the loop's bandwidth. Moreover, the selection of parameter k_p is a tradeoff between loop's bandwidth and dynamic response.

3.5. Time-domain simulation results of the ADALINE-PLL

Figs.18-19 show the time-domain simulation results of the single phase version of the proposed ADALINE-PLL under different control parameters. The grid voltage is assumed to
contain 0.1 p.u. 5th order harmonic and 0.1 p.u. 7th order harmonic components and a transient voltage sag occurs at t=0.05s to test the dynamic response of the ADALINE-PLL. Fig.18
shows the performance of the single-phase ADALINE-PLL with the variation of learning
rate (μ) when the loop regulator gains are selected as:k_p=300, k_i=10000. It can be observed
that if the learning rate is selected too small, the estimation error of the ADALINE-PLL
would be remarkable and there would be significant oscillation in the estimated frequency
and the phase estimation error (see the dash line and the dash dot line in Fig.18). The solid
line in Fig.18 shows the performance of the ADALINE-PLL corresponding to the optimal
learning rate μ=0.035.

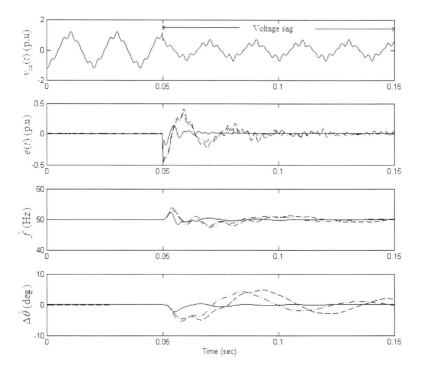

Figure 18. The performance of single-phase ADALINE-PLL with the variation of learning rate (μ) when k_p=300,
k_i=10000. (Solid line: μ=0.035; dash line: μ=0.015; dash dot line: μ=0.025.)

---

2 Jitter — The time variation of a characteristic of a periodic signal in electronics and telecommunications, often in relation to a reference clock source.

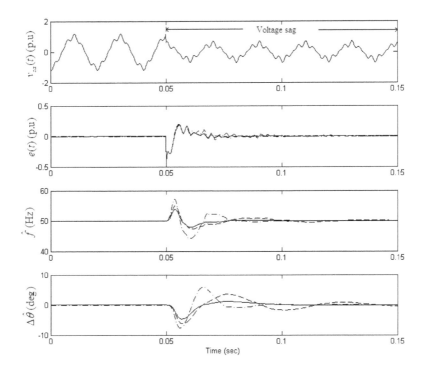

Figure 19. The performance of single-phase ADALINE-PLL with variation of k_p, k_i when μ=0.035. (Solid line:k_p=300, k_i=10000; dash line: k_p=250, k_i=30000; dash dot line: k_p=500, k_i=6000.)

Fig.19 shows the performance of the ADALINE-PLL with the variation of regulator gains when the learning rate is predefined. It can be observed that the dynamic response of the ADALINE-PLL is mainly determined by the proportional gain k_p, if k_p is selected too small, the ADALINE-PLL becomes sluggish and the estimated frequency and phase error decays slowly (dash line in Fig.19). On the other hand, if the gain is selected too high, there would be large overshoot in the estimated frequency and the phase estimation error (the dash dot line in Fig.19). It should be noted that the performance of the ADALINE-PLL is less sensitive to the integration gain k_i. The solid line in Fig.19 shows the performance of ADALINE-PLL corresponding to the optimal regulator parameters.

4. Performance comparison with the existing PLL algorithms

This section presents the performance comparison among the existing PLL algorithms and the proposed ADALINE-PLL. Firstly, a brief introduction of the enhanced PLL (EPLL) and the *park*-PLL is presented. Then, the simulation results of these algorithms are compared with those of the ADALINE-PLL under grid voltage disturbances, such as grid voltage sag, harmonics and random noise contamination scenarios.

4.1. The enhanced phase-locked loop (EPLL)

In recent literature, the enhanced PLL (EPLL) system was proposed (Karimi-Ghartemani et al, 2004). The major improvement introduced by the EPLL is in the PD mechanism, which is replaced by a new strategy allowing more flexibility and provides more information such as amplitude and phase angle. The mechanism of this EPLL is based on estimating in-phase and quadrature-phase amplitudes of the desired signal, hence, has potential application in communication systems which employ quadrature modulation techniques.

The Matlab/Simulink diagram of this EPLL is shown in Fig.20. It can be observed that there are three gains, denoted as k_g, k_p and k_i, which are selected to control the convergence speed for the amplitude, phase and frequency of the fundamental component of the input signal. The guideline for the selection of these gains, however, is not that trivial. The control loop interaction exists since the amplitude, phase and frequency estimation are competing with each other, if any of these gains is varied, it would affect the performance and stability of the closed-loop algorithm. Generally, the gain for the frequency estimation (k_i) should be very small to ensure stability. However, it would result in slow dynamic performance under frequency deviation in the grid voltage. If the frequency estimation is disabled by setting k_i to be zero, steady state error may appear or the algorithm may even diverge under large deviations in the input. Therefore, this EPLL scheme is difficult to be practically implemented, especially for the grid-connected converters which has demanding requirements for tracking accuracy, stability and reliability of the synchronization algorithm (Karimi-Ghartemani et al, 2004).

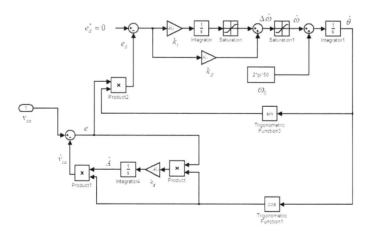

Figure 20. The Matlab/Simulink diagram for the enhanced PLL (EPLL).

4.2. The *Park* phase-locked loop (*Park*-PLL)

The *park*-PLL was another single-phase version of the three-phase synchronous reference frame (SRF) PLL (Filho, R. M. S., et al., 2008). As shown in Fig.21, the circuit diagram of the *park*-PLL consists of two matrix transformations, namely, the Park's transformation and the inverse Park's transformation. The component v_β of the stationary frame is obtained by inverse Park's transformation of the filtered synchronous components v_d' and v_q' in order to emulate a three-phase balanced electric system. The time constants τ_d and τ_q of the two first-order low pass filters (FOLPFs) determines the dynamic characteristics of the phase detection (PD) section.

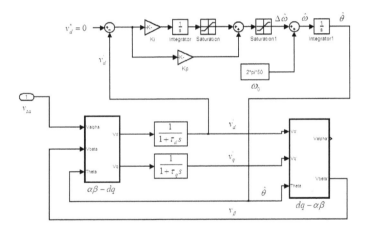

Figure 21. The Matlab/Simulink diagram of the *park*-PLL

It was reported that the PD is always asymptotically stable around the equilibrium condition $\hat\omega \cong \omega$. As for the selection of time constants, if τ_d(or τ_q) is made too small, a pair of real poles will take place and results in a slow dynamic response. On the other hand, if τ_d(or τ_q) is made too high, a pair of complex conjugate poles with small real part will take place, which makes the *park*-PLL slow and oscillatory. It was suggested that the filter cutoff frequency should be equal to about two times line frequency to ensure a fast dynamic response (Filho, R. M. S., et al., 2008).

After the cutoff frequency of the low pass filters is selected, the compensator gains, namely, k_p and k_i, can be set in order to meet dynamic response and line disturbance rejection specifications. However, it should be noted that each harmonic component of order h and amplitude V_h in input grid voltage will produce two components of orders $h\pm1$ and amplitude of $V_h/2$ in the PD output. Besides, a dc component in input voltage will also lead to a fundamental frequency oscillation in the dq components. Therefore, a tradeoff between speed of dynamic response and harmonic rejection capabilities should be achieved to optimize the performance of the $park$-PLL.

4.3. The performance evalution among the EPLL, the $Park$-PLL, and the ADALINE-PLL

Fig.22 shows the simulation results corresponding to the estimated frequency in grid voltage and the phase estimation error when the grid is subjected to 0.7 per unit (p.u.) voltage sag. Here the existing grid synchronization schemes, namely, the enhance PLL (EPLL) and the $park$-PLL are also simulated for the sake of comparison. It can be observed that the $park$-PLL and the EPLL have similar dynamic response in the estimated frequency, with an overshoot of 5Hz when voltage sag occurs. It is interesting to notice that the response time of $park$-PLL and the EPLL is longer when the grid voltage recovers to normal condition. The proposed ADALINE-PLL shows the lowest frequency overshoot compared with other grid synchronization schemes. As far as the phase estimation error is concerned, the phase estimation error of the $park$-PLL and EPLL has high transient overshoot with noticeable oscillations. Whereas, the proposed ADALINE-PLL shows the best dynamic response with smallest phase estimation error with overshoot of about 2 degrees. It can be concluded from the estimated frequency and the phase estimation error that the ADALINE-PLL provides a more robust performance when subject to significant sag in the grid voltage.

Fig.23 shows the simulation results corresponding to the estimated frequency in grid voltage and the phase estimation error when the grid is contaminated by harmonics. The 0.3 per unit (p.u.) 5th order harmonic and 0.3 per unit (p.u.) 7th order harmonic components are added to the grid voltage at t=0.05s with a duration of 0.15s to test the immunity of the various grid synchronization schemes. The $park$-PLL and the EPLL show noticeable oscillations in the estimated frequency when the harmonics are added to the grid voltage. Besides, the $park$-PLL shows longer settling time when the grid voltage recovers to the normal condition. The EPLL shows the highest estimation error in grid frequency with amplitude of about 20 Hz, and the $park$-PLL shows the estimation error of about 10Hz when the harmonics are imposed. However, the proposed ADALINE-PLL shows the lowest frequency overshoot (0.5Hz) and highest estimation accuracy in the estimated frequency compared to the other grid synchronization schemes. Furthermore, the phase estimation error of the $park$-PLL and the EPLL is remarkable during transients, and the $park$-PLL is found to have a large settling time when the grid voltage recovers. Besides, it shows that the EPLL has significant ripples in the phase estimation error. However, the proposed ADALINE-PLL shows negligible estimation error compared to the other algorithms, which implies that the proposed ADALINE-PLL shows better robustness under harmonic contamination in grid voltages.

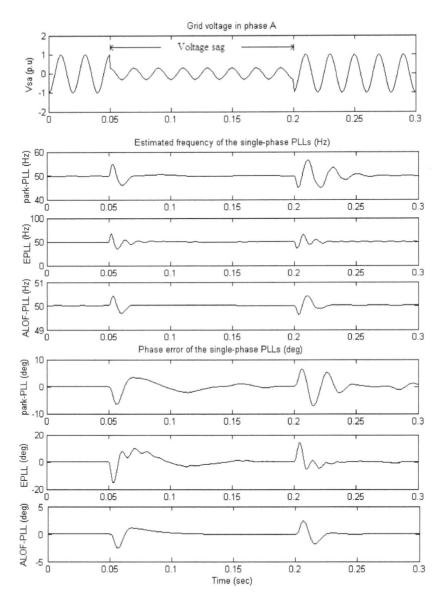

Figure 22. Performance comparison among the EPLL, the park-PLL and the proposed ALOF-PLL algorithm under 0.7 p.u. voltage sag in grid voltages (note: the ADALINE-PLL is abbreviated by ALOF-PLL)

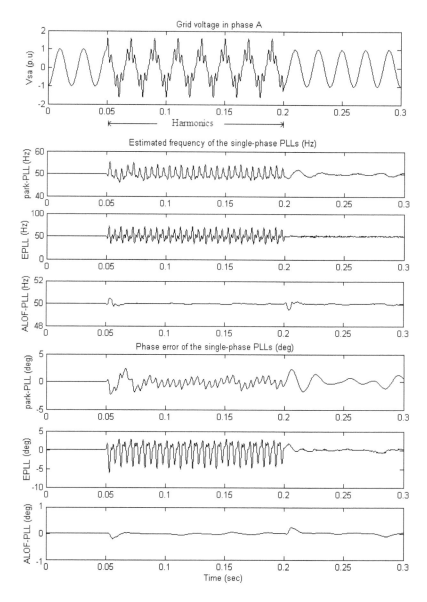

Figure 23. Performance comparison among the *park*-PLL, the EPLL and the proposed ADALINE-PLL algorithm under 0.3 p.u. 5th order harmonic (negative sequence) and 0.3 p.u. 7th order harmonic (positive sequence) components in grid voltages

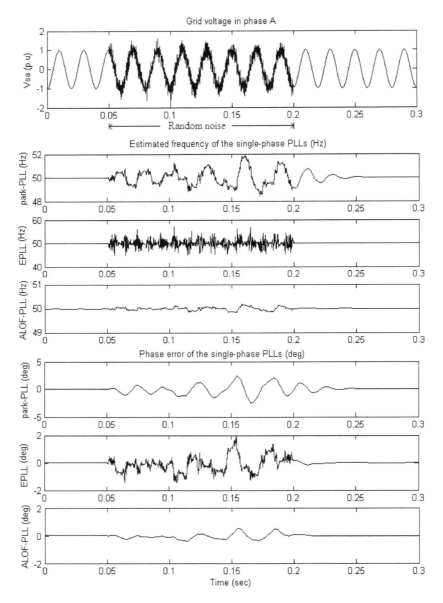

Figure 24. Performance comparison among the *park*-PLL, the EPLL and the ADALINE-PLL algorithm when random noise (power=5e-6) is suddenly applied in grid voltages

Fig.24 shows the simulation results corresponding to the estimated frequency in grid volt-age and the phase estimation error when the grid voltage is contaminated by random noise. The random noise of power density 10e-5 per unit (p.u.) is added to the grid volt-age at t=0.05s with a duration of 0.15s to test the immunity of the various grid synchroni-zation schemes. Similar to the case of a sudden applying harmonics, the park-PLL and EPLL show noticeable oscillations in the estimated frequency when the noise is added to the grid voltage. Besides, the park-PLL shows longer settling time when the grid voltage recovers to the normal condition. The EPLL shows the highest estimation error in grid frequency with amplitude of about 5 Hz, and the park-PLL shows the estimation error of about 2Hz when the noise is imposed. However, the proposed ADALINE-PLL shows the lowest frequency oscillation (0.2Hz) and highest estimation accuracy in the estimated fre-quency compared to the other grid synchronization schemes. Moreover, the phase estima-tion error of the park-PLL and the EPLL is remarkable during transients, and the park-PLL is found to have a large settling time when the grid voltage recovers. Besides, it shows that the park-PLL has the maximum phase estimation error of about 3 degrees, and the phase estimation error of EPLL is less than 2 degrees. However, the proposed ADALINE-PLL shows negligible estimation error compared to the other algorithms, with amplitude of less than 0.5 degree. The estimated frequency and the phase estimation er-ror in Fig.24 indicate that the proposed ADALINE-PLL shows better robustness when grid voltage is contaminated by random noise.

5. Conclusions

The electrical power systems are under a transition to the smart grid owing to the advance-ment of modern control, communication technologies and the requirement of real-time mar-keting. In the smart grid, the power converters are indispensable components which connect the renewable energy resources and the FACTS devices, power quality conditioning devices to the grid. Hence the accurate grid-synchronization of these power converters to the grid is crucial to ensure their stable operation. This book chapter aims to provide a systematic ap-proach for the adaptive linear neural network (ADALINE) algorithm for the real-time har-monic estimation and phase synchronization for the grid-connected converters, which are the fundamental building blocks for the smart grid infrastructure.

The mathematical derivation of the ADALINE algorithm and the ADALINE-PLL scheme is presented, followed by the stability analysis, the continuous domain and the discrete do-main models, and the guidelines for parameter selection of the ADALINE-PLL algorithm. The performance of the ADALINE-PLL is further validated by performance comparison with the existing park-PLL and EPLL algorithms. It can be expected that the presented ADALINE-based algorithms can find wide application in the grid-connected converters for smart grid applications.

Acknowledgment

This work is financially supported by the Fundamental Research Funds for the Central Universities of China under grant No.ZYGX2011J093.

Author details

Yang Han[*]

Address all correspondence to: hanyang_facts@hotmail.com

Dept. of Power Electronics, School of Mechatronics Engineering, University of Electronic Science and Technology of China, Chengdu, China

References

[1] Abdeslam, D. O., Wira, P., Chapuis, Y., & , A. (2007). A unified artificial neural network architecture for active power filters,*IEEE Transactionson Industrial Electronics*, 0278-0046, 54(1), 61-76.

[2] Chang, G. W., Chen, C. I., Liang, Q. W., & (2009, . (2009). A two-stage ADALINE for harmonics and inter-harmonics measurement, *IEEE Transactionson Industrial Electronics*, 0278-0046, 56(6), 2220-2228.

[3] Chang, G. W., Liu, Y. J., & Su, H. J. (2010). On real-time simulation for harmonic and flicker assessment of an industrial system with bulk nonlinear loads,*IEEE Transactionson Industrial Electronics*, Sept. 2010., 0278-0046, 57(9), 2998-3009.

[4] Cirrincione, M., Pucci, M., Vitale, G., & (2008, . (2008). A single-phase DG generation unit with shunt active power filter capability by adaptive neural filtering. *IEEE Transactionson Industrial Electronics,, , 55*(5), 2093-2110, 0278-0046.

[5] Filho, R. M. S., Seixas, P. F. ., Cortizo, P. C., Torres, L. A. B., & Souza, A. F. (2008). Comparison of three single-phase PLL algorithms for UPS applications. *IEEE Transactionson Industrial Electronics*, 0278-0046, 55(8), 2923-2932.

[6] Froehlich, J., Larson, E., & Patel, S. N. (2011). Disaggregated End-Use energy sensing for the smart grid. *IEEE Pervasive Computing*, 10(1), 28-39, 1536-1268.

[7] Han, Y., Khan, M. M., & Chen, C. (2008). A novel harmonic-free power factor corrector based on T-type APF with adaptive linear neural network (ADALINE) control. *Simulation Modeling Practice&Theory*,0156-9190X, 16(9), 1215-1238.

[8]   Han, Y., Xu, L., & Chen, C. (2009). A novel synchronization scheme for grid-connected converters by using adaptive linear optimal filter based PLL (ALOF-PLL). *Simulation Modeling Practice and Theory*, 0156-9190X, 17(7), 1299-1345.

[9]   Jarventausta, P., Repo, S., & Partanen, J. (2010). Smart grid power system control in
distributed generation environment. *Annual Reviews in Control*, 34(2), 277-286,
1367-5788.

[10] Karimi-Ghartemani, M. ., & Iravani, M. R. (2004). A method for synchronization of
power electronic converters in polluted and variable-frequency environments. *IEEE
Transactions on Power Systems*, 19(3), 1263-1270, 0885-8950.

[11] Kefalas, T. D., Kladas, A. G., & (2010, . (2010). Harmonic impact on distribution transformer no-load loss. *IEEE Transactionson Industrial Electronics*, 0278-0046, 57(1),
193-200.

[12] Simon H., Adaptive Filter Theory, Prentice Hall, Fourth Edition,. (2002). 13978.

[13] Sauter, T., & Lobashov, M. (2011). End-to-End communication architecture for smart
grids, *IEEE Transactions on Industrial Electronics*, 0278-0046, 58(4), 1218-1228.

[14] Varaiya, P. P., Wu, F. F., & Bialek, J. W. (2011). Smart operation of smart grid: Risk-
limiting dispatch. Proceedings of the IEEE, 0018-9219, 99(1), 40-57.

[15] Wira, P., Abdeslam, D., & Mercklé, J. (2010). Artificial Neural Networks to Improve
Current Harmonics Identification and Compensation. in: *Intelligent Industrial Systems:
Modelling, Automation and Adaptive Behavior* (Editor: G. Rigatos), IGI Publications.
978-1-61520-849-4

[16] Yin, J. J., Tang, W., & Man, K. F. (2010). A comparison of optimization algorithms for
biological neural network identification. *IEEE Transactionson Industrial Electronics*,
0278-0046, 57(3), 1127-1131.

Chapter 4

Adaptive Analysis of Diastolic Murmurs for Coronary Artery Disease Based on Empirical Mode Decomposition

Zhidong Zhao, Yi Luo, Fangqin Ren, Li Zhang and Changchun Shi

Additional information is available at the end of the chapter

1. Introduction

Coronary Artery Disease (CAD) is a leading type of heart disease in the world caused by the gradual build-up of plaque on the walls of the arteries. Due to CAD's high incidence rate and mortality, it is very harmful to human health. CAD can develop slowly and silently over years without any symptoms. Early diagnose of CAD is one of the most important medical research areas. Diastolic murmurs that occur as additional components in the heart sound signal provide clinicians with valuable diagnostic and prognostic information about the function of heart valves. When coronary arteries become narrowed or blocked, the turbulence appears which is produced by blood moving across the stenotic arteries. During the relatively quiet diastolic period of the cardiac cycle, the murmurs are likely to be loudest when coronary blood flow is maximal. Initial studies show that diastolic murmurs produced by coronary arterial stenosis contain higher frequency components.

The heart sound signal represents the mechanical activity of the cardiohemic system, which is complicated and non-stationary. It contains physiological and pathological information between the heart and the various parts of the body, so it can be used in diagnosis of heart disease. Heart sound has been widely used in diagnosis of heart disease and many methods have been adopted to aid the diagnosis [1, 2]. The heart sound signal generally can be separated into four parts: the 1st heart sound S1, the systolic period, the 2nd heart sound S2 and the diastolic period, shown in figure 1.

Diastolic murmurs occur between S2 and the next S1 when the heart muscle relaxes between beats. Heart murmurs are usually considered pathological. They can be caused by some kinds of heart attacks, such as coronary artery stenosis, aortic regurgitation, etc. Diastolic murmurs can provide clinicians with valuable diagnostic and prognostic information about the function of heart valves.

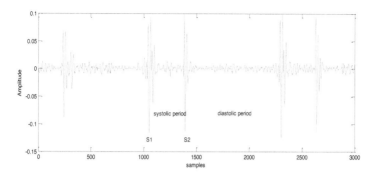

Figure 1. Heart sound signal

Short Time Fourier Transform, Wigner-Ville Distribution and Wavelet Transform, etc., have some inherent limitations [3, 4, 5]. Short Time Fourier Transform involves an intrinsic trade-off between time resolution and frequency resolution. In Wigner-Ville distribution, the inherent cross-term interferences often mask the true time-frequency information associated with the signal of interest. The wavelet transform has received considerable attention in recent years. It provides a multi-resolution representation of signals, however, it is not adaptive in nature; once the wavelet mother function is given, one will have to use it to analyse all the data. In addition, the wavelet transform also underlies an uncertainty principle. In 1998, Dr.Norden Huang proposed a novel signal processing algorithm: the Hilbert Huang Trans-form (HHT) [6, 7]. It has proved to be a powerful tool to analyse non-stationary and nonlinear signals. The key parts of HHT are the Empirical Mode Decomposition (EMD) and Hilbert transform. EMD can decompose adaptively diastolic murmurs into a finite and usually small number of Intrinsic Mode Functions (IMFs) that admit a well-behaved Hilbert transform. The Hilbert transform of IMFs can yield instantaneous frequency and instantaneous amplitude. The local energy and instantaneous frequency derived from the IMFs give the fine-resolution frequency-time distribution of the energy that is designated as the Hilbert spectrum. The three-dimensional distribution can reflect the inherent essential characteristic of the signal.

The paper is organized as follows: section 2 introduces generalized wavelet shrinkage denoising method. In section 3, the Hilbert spectrum based on EMD and marginal spectrum distributions of diastolic murmurs are studied; a new method to restrict the end effect of EMD is proposed in section 4.In section 5, the algorithm based on the Empirical Mode Decomposi-tion (EMD) and Teager Energy Operation (TEO) is proposed as an effective approach for estimating the instantaneous frequency of diastolic murmurs. Finally, some conclusions are given in section 6.

2. Wavelet shrinkage method

We consider the following model of a discrete noisy signal:

$$x = \theta + \sigma z \tag{1}$$

The vector x represents noisy signal and θ is an unknown original clean signal. z is independent identity distribution Gaussian white noise with mean zero and unit variance. σ is intensity of noise. For simplicity, we assume intensity of noise is one.

The step of wavelet shrinkage is defined as follows:

1. Apply discrete wavelet transform to observe noisy signals.

2. Estimate noise and threshold value, thresholding the wavelet coefficients of the observed signal.

3. Apply the inverse discrete wavelet transform to reconstruct the signal.

The wavelet shrinkage method relies on the basic idea that the energy of signal will often be concentrated in a few coefficients in the wavelet domain while the energy of noise is spread among all coefficients in the wavelet domain. Therefore, the nonlinear shrinkage function in the wavelet domain will tend to keep a few larger coefficients over threshold value that represent the signal, while noise coefficients' down threshold value will tend to reduce to zero.

In wavelet shrinkage, how to select the threshold function and how to select the threshold value are most crucial. Donohue introduced two kinds of thresholding functions: 'hard threshold function' and 'soft threshold function'.

$$\delta_\lambda^H(x) = \begin{cases} 0 & |x| \le \lambda \\ x & |x| > \lambda \end{cases} \tag{2}$$

$$\delta_\lambda^S(x) = \begin{cases} 0 & |x| \le \lambda \\ x - \lambda & x > \lambda \\ x + \lambda & x < -\lambda \end{cases} \tag{3}$$

The hard threshold function (2) results in larger variance and can be unstable because of the discontinuous function. The soft threshold function (3) results in unnecessary bias due to shrinkage of the large coefficients to zero. We build the generalized threshold function:

$$\delta_\lambda^m(x) = x - \frac{\lambda^m}{x^{m-1}} \quad m = 1\,2\infty \tag{4}$$

λ is threshold value.

When m is an even number:

$$\delta_\lambda^m(x) = x - xI(|x| \le \lambda) - \frac{\lambda^m}{x^{m-1}}I(|x| > \lambda) \tag{5}$$

When m is odd number:

$$\delta_\lambda^m(x) = x - xI(|x| \le \lambda) - \frac{\lambda^m}{x^{m-1}}I(|x| > \lambda)sign(x) \tag{6}$$

When m=1, it is the soft threshold function; when m= ∞, it is the hard threshold function. When m=2 it is Non-Negative Garrote threshold function. We show slope signal as an example, Figure2 illustrates the generalized threshold functions for different m.

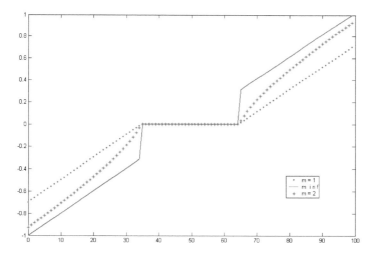

Figure 2. Generalized threshold function

It can clearly be seen that when the coefficient is small, the smaller m is, the closer the generalized function is to the soft threshold function; when the coefficient is big, the bigger m is, the closer the generalized function is to the hard threshold function; when m lies between 1 and ∞, the general threshold function achieves a compromise between hard and soft threshold function. With careful selection of m, we can achieve better denoising performance [8, 9].

We derived the exact formula of mean, bias, variance and l_2 risk for the generalized threshold function.

Let $x \sim N(\theta, 1)$

$$A_m(\theta) = \int_\lambda^\infty \frac{\phi(x-\theta) - \phi(x+\theta)}{x^m} dx \, B_m(\theta) = \int_\lambda^\infty \frac{\phi(x-\theta) + \phi(x+\theta)}{x^m} dx$$

ϕ And Φ are density and probability function of standard Gaussian random variable respectively. Then:

Mean:

$$M^m(\lambda, \theta) = M^H(\lambda, \theta) - \lambda^m A_{m-1}(\theta) \tag{7}$$

Bias:

$$SB^m(\lambda, \theta) = (M^m(\lambda, \theta) - \theta)^2 \tag{8}$$

Variance:

$$V^m(\lambda, \theta) = V^H(\lambda, \theta) - 2\lambda^m B_{m-2}(\theta) - \lambda^{2m} A_{m-1}^2(\theta) + \lambda^{2m} B_{2m-2}(\theta) + 2\lambda^m M^H(\lambda, \theta) A_{m-1}(\theta) \tag{9}$$

l2 Risk:

$$\rho_\lambda^m(\theta) = E(\delta_\lambda^m(x) - \theta)^2 = \rho_\lambda^H(\theta) - 2\lambda^m B_{m-2}(\theta) + \lambda^{2m} B_{2m-2}(\theta) + 2\theta\lambda^m A_{m-1}(\theta) \tag{10}$$

Where

$$\rho_\lambda^m(\theta) = E(\delta_\lambda^m(x) - \theta)^2 = \rho_\lambda^H(\theta) - 2\lambda^m B_{m-2}(\theta) + \lambda^{2m} B_{2m-2}(\theta) + 2\theta\lambda^m A_{m-1}(\theta)$$

$$M^H(\lambda, \theta) = \theta + \theta[1 - \Phi(\lambda - \theta) - \Phi(\lambda + \theta)] + \phi(\lambda - \theta) - \phi(\lambda + \theta)$$

$$V^H(\lambda, \theta) = (\theta^2 + 1)(2 - \Phi(\lambda - \theta) - \Phi(\lambda + \theta)) + (\lambda + \theta)\phi(\lambda - \theta) + (\lambda - \theta)\phi(\lambda + \theta) - M^H(\frac{\lambda\theta2}{,},$$

$$\rho_\lambda^H(\theta) = 1 + (\theta^2 - 1)(\Phi(\lambda - \theta) - \Phi(-\lambda - \theta)) + (\lambda + \theta)\phi(\lambda + \theta) + (\lambda - \theta)\phi(\lambda - \theta)$$

$M^m(\lambda, \theta)$, $SB^m(\lambda, \theta)$, $V^m(\lambda, \theta)$ are the mean, bias, variance and risk of generalized threshold function. When m is 1, 2, $\rho_\lambda^m(\theta)$, they are the mean, bias, variance and risk of the risk of soft, Non-Negative Garrote, hard threshold functions, respectively.

The soft threshold function provides smoother results in comparison with the hard threshold function; however, the hard threshold function provides better edge preservation in comparison with the soft threshold function. The hard threshold function is discontinuous and this leads to the oscillation of denoised signal. The soft threshold function tends to have bigger bias because of shrinkage, whereas the hard threshold function tends to have bigger variance because of discontinuity. The Non-Negative Garrote threshold function is the trade-off between the hard and soft threshold function. Firstly, it is continuous; secondly, the shrinkage amplitude is smaller than the soft threshold function.

Stein Unbiased Risk Estimate (SURE) [10] is an adaptive threshold selection rule which is data driven. The threshold value minimizes an estimate of the risk.

If ∞ is weakly differentiable, for single coefficient, $\hat{\theta}_k = x_k + H(x_k)$, $k = 1...N$, H is true risk. Then

$$\hat{\rho}(x_k, \lambda) = 1 + 2(\frac{d}{dx_k} H(x_k)) + H^2(x_k) \tag{11}$$

is the unbiased risk estimate of $\hat{\rho}(x_k, \lambda) = 1 + 2(\frac{d}{dx_k} H(x_k)) + H^2(x_k)$:

Proof:

$$\rho(x_k, \lambda)E\| \hat{\theta}_k - \theta_k \|^2 = E(x_k + H(x_k) - \theta_k)^2 = E(z_k + H(x_k))^2 = 1 + 2E(z_k H(x_k)) + E(H^2(x_k))$$

Where $= 1 + 2E(z_k H(\theta_k + z_k)) + E(H^2(x_k))$ and by partial integral

$z_k = x_k - \theta_k$

Then

$$E(z_k H(\theta_k + z_k)) = \frac{1}{\sqrt{2\pi}} \int z_k H(\theta_k + z_k) e^{\frac{\xi^2}{2}} d\xi = \frac{1}{\sqrt{2\pi}} \int (\eta_k - \theta_k) H(\eta_k) \exp(-\frac{(\eta_k - \theta_k)^2}{2}) d\eta_k$$

$$= \frac{1}{\sqrt{2\pi}} \int \exp(-\frac{(\eta_k - \theta_k)^2}{2}) \frac{dH(\eta_k)}{d\eta_k} d\eta_k = E(\frac{dH(\eta_k)}{d\eta_k} \mid \eta_k = x_k)$$

So

$$\rho(x_k, \lambda) = E\| \hat{\theta}_k - \theta_k \|^2 = 1 + 2E(\frac{dH(x_k)}{dx_k}) + E(H^2(x_k)) \text{ is the unbiased risk estimate of true risk}$$

$$E[\hat{\rho}(x_k, \lambda)] = 1 + 2(\frac{dH(x_k)}{dx_k}) + H^2(x_k).$$

For the generalized threshold function (5) and single coefficient, when m is even,

$\rho(x_k, \lambda)$

The SURE is

$$SURE(x_k, \lambda) = 1 + (x_k^2 - 2)I(|x_k| \le \lambda) + (\frac{2(m-1)\lambda^m}{x_k^m} + \frac{\lambda^{2m}}{x_k^{2m-2}})I(|x_k| > \lambda) \tag{12}$$

When m is odd,

$$SURE(x_k, \lambda) = 1 + (x_k^2 - 2)I(|x_k| \le \lambda) + (\frac{2(m-1)\lambda^m}{x_k^m} + \frac{\lambda^{2m}}{x_k^{2m-2}})I(|x_k| > \lambda)$$

The SURE is

$$SURE(x_k, \lambda) = 1 + (x_k^2 - 2)I(|x_k| \le \lambda) + \frac{\lambda^{2m}}{x_k^{2m-2}}I(|x_k| > \lambda) + \frac{2(m-1)\lambda^m}{x_k^m}I(x_k > \lambda) - \frac{2(m-1)\lambda^m}{x_k^m}I(x_k < -\lambda) \tag{13}$$

Suppose   wavelet   coefficients   are

$$SURE(x_k, \lambda) = 1 + (x_k^2 - 2)I(|x_k| \le \lambda) + \frac{\lambda^{2m}}{x_k^{2m-2}}I(|x_k| > \lambda) + \frac{2(m-1)\lambda^m}{x_k^m}I(x_k > \lambda) - \frac{2(m-1)\lambda^m}{x_k^m}I(x_k < -\lambda)$$

, the threshold value λ is set to minimize the estimate of the $x_1....x_N$ risk for the given data,

$$\lambda = \arg\min_{\lambda \ge 0} \sum_{k=1}^{N} SURE(x_k, \lambda) \tag{14}$$

For hard threshold function (m is ∞), because H (x) is discontinuity, the SURE is illogical.

The noisy PCG signal is processed using the method mentioned above. For the generalized threshold functions, parameter m is selected as 2 which is simple and provides a good compromise between the hard and soft threshold function. The data-driven SURE threshold value is used. The filtered PCG signal is illustrated as figure 4(a). The phase space diagram of the filtered PCG signal is shown in figure 4(b). From visual inspection of figure 3, the PCG signal is much cleaner after being denoised; the first heart sound, the systolic period, the second heart sound and the diastolic period can be clearly identified. The results indicate that the method we have proposed significantly reduces noise and preserves well the characteristics of the PCG signal.

3. Analysis of diastolic murmurs for coronary artery disease based on empirical mode decomposition

Since a novel signal processing algorithm - the Hilbert HuangTransform (HHT) - was proposed by N.E.Huang in 1998 [6], it has been seen as a data-driven tool for nonlinear and non-stationary signal processing. HHT consists of two parts: the EMD and Hilbert transform. EMD as the important part of the HHT that can adaptively decompose signal into a finite and often a series of small numbers of Intrinsic Mode Functions (IMFs) subjected to the following two conditions:

1.   In the whole dataset, the number of extrema and the number of zero-crossing must either be equal or differ at most by one.

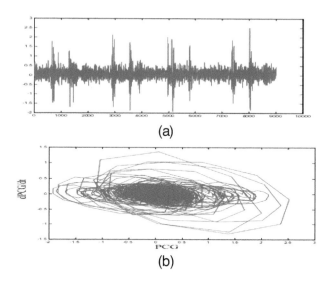

Figure 3. a) Noisy PCG signal (b) Phase space diagram of the noisy signal

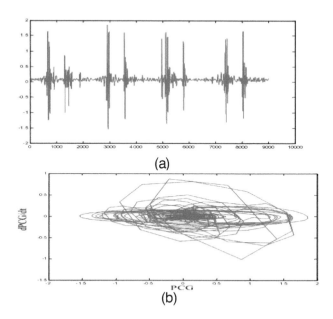

Figure 4. a) PCG signal after denoising (b) Phase space diagram of denoised signal

2.   At any time, the mean value of the envelope of the local maxima and the envelope of the
      local minima must be zero.

These two conditions guarantee the well-behaved Hilbert transform. The IMFs represent the
oscillatory modes embedded in the signal. Most signals include more than one oscillatory
mode and are not IMFs. EMD is a numerical sifting process to decompose a signal into a finite
number of hidden fundamental intrinsic oscillatory modes, i.e., IMFs. Applying the Hilbert
transform to each IMF, the instantaneous frequency and amplitude of each IMF can be obtained
which constitute the time-frequency-energy distribution of the signal, called the Hilbert
spectrum. The Hilbert spectrum provides higher resolution and concentration in the time-
frequency plane, and avoids the false high frequency and energy dispersion existent in the
Fourier spectrum.

Figure5 shows a classical IMF. The IMFs represent the oscillatory modes embedded in the
signal. Each IMF actually is a zero mean monocomponents AM-FM signal with the following
form:

$$x(t) = a(t)\cos\phi(t) \tag{15}$$

with time varying amplitude envelope $x(t) = a(t)\cos\phi(t)$ and phase $a(t)$. The amplitude and
phase both have physical and mathematical meaning.

Most signals include more than one oscillatory mode, so they are not IMFs. EMD is a numerical
sifting process to disintegrate empirically a signal into a finite number of hidden fundamental
intrinsic oscillatory modes, that is, IMFs. The sifting process can be separated into the following
steps:

1.   Finding all the local extrema, including maxima and minima; then connecting all the
      maxima and minima of signal x(t) using smooth cubic splines to get its upper envelope
      $\phi(t)$ and lower envelope $x_{up}(t)$.

2.   Subtracting mean of these two envelopes $x_{low}(t)$ from the signal to get their difference:
      $m_1(t) = (x_{up}(t) + x_{low}(t))/2$.

3.   Regarding the $h_1(t) = x(t) - m_1(t)$ as the new data and repeating steps 1 and 2 until the
      resulting signal meets the two criteria of an IMF, defined as $h_1(t)$. The first IMF $c_1(t)$
      contains the highest frequency component of the signal. The residual signal $c_1(t)$ is given
      by $r_1(t)$.

4.   Regarding $r_1(t) = x(t) - c_1(t)$ as new data and repeating steps (1) (2) (3) until extracting all
      the IMFs. The sifting procedure is terminated until the m-th residue $r_1(t)$ becomes less than
      a predetermined small number or becomes monotonic.

The original signal x (t) can thus be expressed as follows:

$$x(t) = \sum_{j=1}^{M} c_j(t) + r_M(t) \tag{16}$$

$x(t) = \sum_{j=1}^{M} c_j(t) + r_M(t)$ is an IMF where j represents the number of corresponding IMFs and $c_j(t)$ is residue. The EMD decomposes non-stationary signals into narrow-band components with decreasing frequency. The decomposition is complete, almost orthogonal, local and adaptive. All IMFs form a completely and nearly orthogonal basis for the original signal. The basis comes directly from the signal, which guarantees the inherent characteristic of signal and avoids the diffusion and leakage of signal energy. The sifting process eliminates riding waves, so each IMF is more symmetrical and is actually a zero mean AM-FM component.

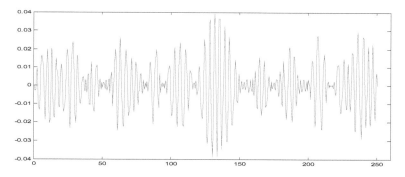

Figure 5. A classical IMF

Heart sounds are recorded from the chest of normal objects and CAD patients using a specially designed high sensitivity cardiac microphone. The ECG signals are also recorded as a time reference to aid in locating the diastolic phase. For each cycle, the central portion of diastole is digitized (sample frequency equals 2.0 kHz).

Figure6 shows the diastolic murmurs of a normal object. Figure7 shows the IMFs of the murmur obtained by EMD. The diastolic murmurs can be decomposed into six IMFs. The Hilbert spectrum is shown in figure 8. The vertical bars on the right of the panel give the relative amplitude scale. Figure6 provides more distinct information on the time-frequency contents of diastolic murmurs, which reveals clearly the dynamic characteristic of murmurs in the time-frequency plane. The Hilbert spectrum contains no energy with frequency above 350Hz. The spectrum appears in the skeleton form and can provide the frequency variations from one instance to the next. Figure 9 shows the marginal spectrum of the diastolic murmurs. It can be clearly seen that the energy mainly concentrates on the lower frequency domain.

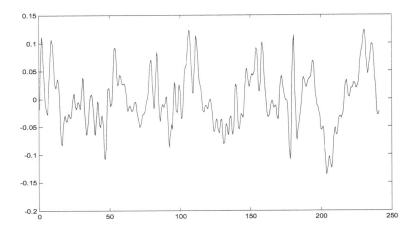

Figure 6. Diastolic murmurs of a normal object

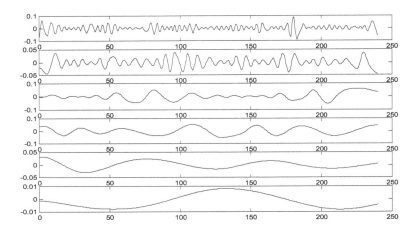

Figure 7. IMFs of diastolic murmurs from the normal people

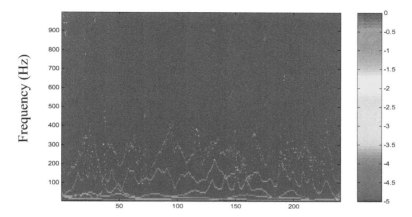

Figure 8. Hilbert spectrum of the diastolic murmurs

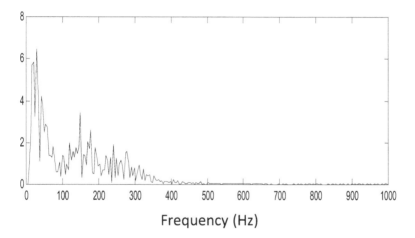

Figure 9. Marginal spectrum of the diastolic murmurs

Figure 10 shows the diastolic murmurs of the CAD patient, as diagnosed by coronary artery radiography. The left anterior descending artery is stenosed about 60% and the right coronary artery is stenosed about 85%. Figure 11 shows the IMFs of the murmur obtained by EMD. The diastolic cardiac cycle can be decomposed into six IMFs. The Hilbert spectrum is illustrated in figure 12. Figure 13 shows the marginal spectrum of diastolic murmurs. The HHT spectrum has superior temporal and frequency resolutions. The spectrums show precise time-frequency representation of signal. The energies spread over a much wider frequency domain. Much higher spectral energies are concentrated on high frequency compared with those of normal people. More energy distributes in the frequency band over 200Hz and a peak also lies around 350Hz, which often does not appear in diastolic murmurs of normal people. It can be explained as follows: for the CAD patient, the narrowed coronary arteries lead to the blood flow in coronary artery changing from laminar flow to turbulence flow, from simplicity to complexity. Coronary arterial stenosis gives rise to high frequencies of diastolic murmurs. The EMD method makes no assumption about the linearity or stationarity of the signal, and the IMFs are usually easy to interpret and relevant to the underlying dynamic processes being studied.

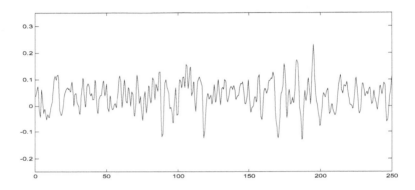

Figure 10. Diastolic Murmurs of CAD patient

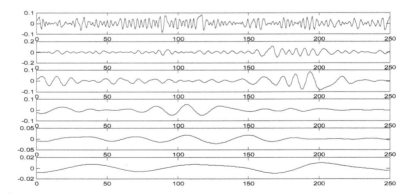

Figure 11. Six IMFs of diastolic murmurs from patient

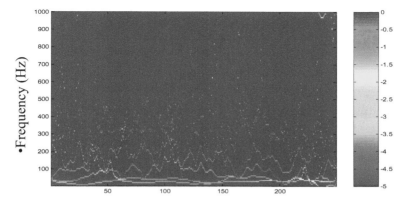

Figure 12. Hilbert spectrum of the diastolic murmurs from patient

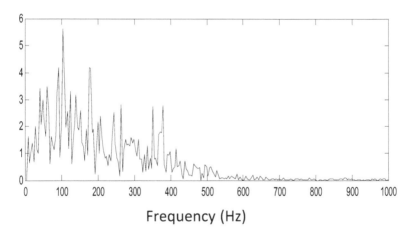

Figure 13. Marginal spectrum of the diastolic murmurs from patient

4. A new method for processing end effect in empirical mode decomposition

In the procedure of EMD, the cubic splines interpolation creates top and bottom envelopes that are implemented in the first step of the above sifting process. It is difficult to interpolate data near the beginnings or ends, where the cubic splines can have swings. The common method to deal with end effect is to consider the end points both as maximum and minimum locations with values unchanged, but this method will give a distorted view of the local mean near the boundaries. We propose a simpler method to restrict the end effect in spline interpolation [11]. The key points are to determine the values and locations of extrema nearby end points. Suppose the length of data x is N, the steps can be implemented as follows:

1. Finding all the maxima and minima, and considering the end points both as maximum and minimum, that is, maximum= [1 maximum N] and minimum= [1 minimum N].

2. The end points are still considered both as maximum and minimum, whereas their values can be adapted to $r_M(t)$ and δ_1, γ_1. Taking δ_N, γ_N, δ_1 as the mean of all maximum except for the first and last maximum (the subscript represents location of maximum). Similarly, taking δ_N, γ_1 as the mean of all minimum except for the first and last minimum (the subscript represents location of minimum).

3. Comparing γ_N with x (1), δ_1 with x (N), δ_N with x(1) and γ_1 with x (N), respectively.

If γ_N<x(1) then δ_1= x (1);if δ_1< x (N) then δ_N= x (N); if δ_N> x (1)then γ_1=x (1);If γ_1>x(N) then γ_N= x(N).

4. Using cubic splines interpolation to get top and bottom envelopes, and repeating the second step of above sifting process to extract IMF.

The performance of the proposed method is compared with the traditional method where the endpoints are considered both as maximum and minimum with values unchanged. As an example, we decompose a sinusoid signal by the sifting process. Figure 14 shows the signal.

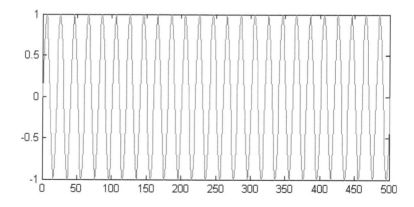

Figure 14. A sinusoid signal

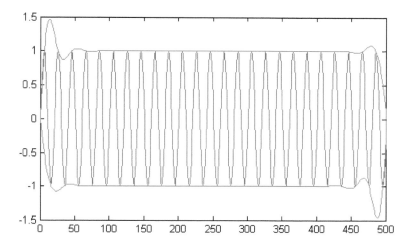

Figure 15. Cubic splines interpolation in sifting process using the traditional method

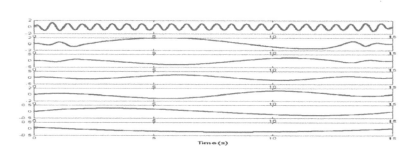

Figure 16. IMFs of the sinusoid signal

Firstly, we consider the endpoints both as maximum and minimum with value unchanged. Figure 15 shows the top and bottom envelopes calculated by cubic splines interpolation in the sifting process. Top and bottom red dash dot line represent the envelopes. The sinusoid signal is decomposed into six IMFs and one residue by sifting process as depicted in figure 16.

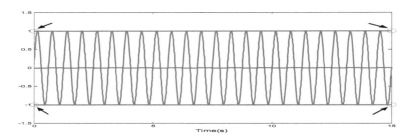

Figure 17. Cubic splines interpolation in sifting process using the proposed method

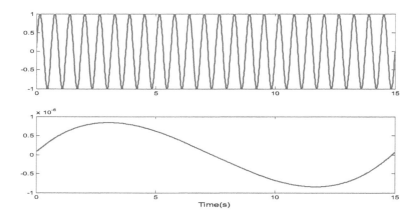

Figure 18. IMF and residue of the sinusoid signal using the proposed method

Secondly, applying the proposed method above to restrict the end effect, figure 17 shows the top and bottom envelopes calculated by cubic splines interpolation in the sifting process. Red circles represent the end values predicted. The sinusoid signal is decomposed into one IMF and a residue by the sifting process as depicted in figure 18. The IMF is just the sinusoid and the value of the residue is smaller than 10^{-6}. From figure 18, it can easily be seen that the swings appear near both ends and propagate inwards and produce superfluous IMFs. Actually, the sinusoid signal is an IMF itself in nature because it satisfies the IMF conditions which has the same numbers of zero-crossing and extrema, and can also be local symmetric. Therefore, the sifting process as represented by figure 18 should extract only one IMF. The results indicate that the method we proposed is effective.

5. Instantaneous frequency estimation of diastolic murmurs based on EMD and TEO

Diastolic murmurs can provide clinicians with valuable diagnostic and prognostic information about the function of heart valves. Quantitative analysis of instantaneous frequency (IF) of the murmurs can aid diagnosis [1, 13].

Instantaneous Frequency (IF) is an important signal characteristic, which characterizes the transients and fast changes in frequency as time progresses. The IF of diastolic murmur is used to describe the time-varying spectral contents of the characteristic frequency bands that are of interest for cardiovascular research. The IF of a signal is traditionally obtained by taking the first derivative of the phase of the signal with respect to time using the Hilbert transform. However, this definition is questionable and will mislead interpretation of instantaneous frequency, such as negative frequency. Instantaneous frequency can also be obtained from a time–frequency distribution (TFD) as the first conditional moment in the frequency, suggesting that the instantaneous frequency is the average frequency at each time, whereas the cross terms existing in TFD will lead to a very rapid degradation of performance and severely pollute the instantaneous frequency estimation [14].

TEO is a powerful nonlinear operator and has been successfully used in a number of applications including speech signal processing, image processing, etc. [15]. TEO can track the modulation energy and estimate the instantaneous amplitude and frequency of AM-FM signals with the form

$$x(t) = a(t)\cos[2\pi \int_0^t \omega(\tau)d\tau] \tag{17}$$

$x(t) = a(t)\cos[2\pi \int_0^t \omega(\tau)d\tau]$ and $a(t)$ are the instantaneous amplitude and frequency respectively.

In continuous time domain, TEO is defined by

$$\Psi(x(t)) = [\dot{x}(t)]^2 - x(t)\ddot{x}(t) \tag{18}$$

$\Psi(x(t)) = [\dot{x}(t)]^2 - x(t)\ddot{x}(t)$ corresponds to continuous signal, $x(t)$ and $\dot{x}(t)$ are the first order and second order time derivatives of $\ddot{x}(t)$ respectively.

For example, for a sinusoid signal $x(t)$, the TEO gives

$$\Psi(x(t)) = A^2\omega^2 \tag{19}$$

For a monochromatic signal, the output by TEO is proportional to the squared product of frequency and amplitude. The TEO of the first order derivative $\Psi(x(t)) = A^2\omega^2$ of $\dot{x}(t)$ produce the output:

$$\Psi(\dot{x}(t)) = A^2\omega^4 \tag{20}$$

The two results above can be combined to estimate the frequency and amplitude of the signal $\Psi(\dot{x}(t)) = A^2\omega^4$ as follows [14]:

$$\hat{\omega}^2(t) = \frac{\Psi(\dot{x}(t))}{\Psi(x(t))} \tag{21}$$

$$|\hat{A}^2(t)| = \frac{\Psi^2(x(t))}{\Psi(\dot{x}(t))} \tag{22}$$

The estimate of instantaneous frequency and amplitude above are also suitable for AM, FM and AM-FM signals.

The discrete-time counterpart of TEO can be defined as:

$$\Psi(x(n)) = x^2(n) - x(n-1)x(n+1) \tag{23}$$

A discrete-time real value AM-FM signal that is usually used to model time-varying amplitude and frequency patterns can be expressed as:

$$x(n) = a(n)\cos(\phi(n)) = a(n)\cos(\omega_c n + \omega_m \int_0^n q(k)dk + \theta) \tag{24}$$

Where $x(n) = a(n)\cos(\phi(n)) = a(n)\cos(\omega_c n + \omega_m \int_0^n q(k)dk + \theta)$ is the time-varying amplitude modulation, $a(n)$ is the carrier frequency, ω_c is the maximum frequency deviation from the carrier frequency and ω_m, $0 < \omega_m < \omega_c$ is the frequency deviation function and $|q(n)| \leq 1$ is the initial phase shift. The derivative of the phase θ, that is, the FM part of the signal is called as instantaneous frequency:

$$\omega(n) = \frac{d\phi(n)}{dn} = \omega_c + \omega_m q(n) \tag{25}$$

The instantaneous frequency $\omega(n) = \dfrac{d\phi(n)}{dn} = \omega_c + \omega_m q(n)$ and amplitude $\omega(n)$ of the AM-FM modulated signal $a(n)$ at any time instant can be respectively demodulated by applying the TEO to $x(n)$ and its difference, which is called the Discrete Energy Separation Algorithm (DESA):

$$y(n) = x(n) - x(n-1) \tag{26}$$

$$\omega(n) = \arccos\left(1 - \frac{\Psi\{y(n)\} + \Psi\{y(n+1)\}}{4\Psi\{x(n)\}}\right) \tag{27}$$

$$|a(n)| = \sqrt{\frac{\Psi\{x(n)\}}{\sin^2(\omega(n))}} \tag{28}$$

or

$$\omega(n) = \frac{1}{2}\arccos\left(1 - \frac{\Psi\{x(n+1) - x(n-1)\}}{2\Psi\{x(n)\}}\right) \tag{29}$$

$$|a(n)| = \frac{2\Psi\{x(n)\}}{\sqrt{\Psi\{x(n+1) - x(n-1)\}}} \tag{30}$$

The estimates above are valid under the assumptions that the signal does not vary too fast nor too much compared to the carrier frequency. In general, the first demodulation algorithm (26) ~ (28) is called DESA-1 where '1' implies the approximation of derivatives with a single sample difference. That is, the signal derivative is approximated by the average of forward and backward 1-point differences. The second demodulation algorithm (29) ~ (30) is called DESA-2 where '2' implies a difference between samples whose time indices differ by 2. Both DESA-1 and DESA-2 algorithms yield very small errors and can give the accurate estimate of instantaneous frequency. The DESA-2 algorithm is less computationally complex and has an excellent, almost instantaneous, time resolution which can also lead to a simpler mathematical analysis. In this paper, we focus on the instantaneous frequency rather than the instantaneous amplitude by DESA-2.

Figure 19 shows an AM-FM signal $\left|a(n)\right| = \dfrac{2\Psi\{x(n)\}}{\sqrt{\Psi\{x(n+1) - x(n-1)\}}}$ where

$$a(n) = 1 + 0.6\cos(0.01\pi n)$$

$$\phi(n) = \frac{\pi}{10}n + \cos\frac{\pi}{80}n \tag{31}$$

The theoretic instantaneous frequency is shown in figure 20. The estimated instantaneous frequency by DESA-2 is shown in figure 21. The estimated amplitude envelope is also illustrated in figure 22. Note that there are no apparent discrepancies between the real values and the DESA-2 calculations. The errors are very slow but less smooth. The results indicate that DESA-2 can be used to track the instantaneous frequency and amplitude accurately.

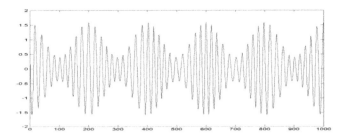

Figure 19. Original AM-FM signal

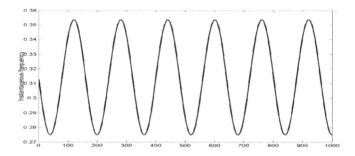

Figure 20. Theoretic instantaneous frequency

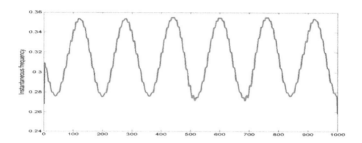

Figure 21. Estimated instantaneous frequency by DESA-2

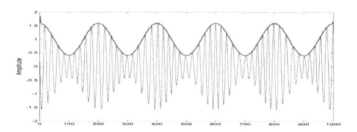

Figure 22. Estimated amplitude envelope by DESA-2

Another mixture signal is composed of two linear swept-frequency signals shown in figure 23. The frequency of one chirp signal varies from 1Hz to 0.1 Hz and the other varies from 2 Hz to 0.1 Hz. The estimated IF is shown in figure 24. The two chirp signals are also better identified and localized except for near boundaries.

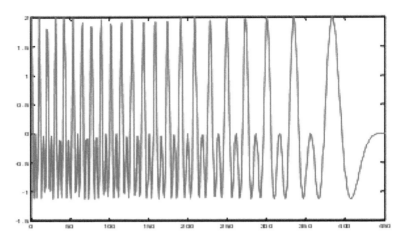

Figure 23. A mixture signal of two chirp signals

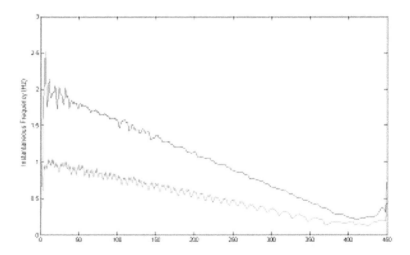

Figure 24. Estimated IF of two IMFs by DESA-2

In this paper, we present a novel method to estimate the IF of diastolic murmurs using Empirical Mode Decomposition (EMD) and nonlinear the Teager Energy Operator (TEO). EMD has been analysed as in section 3 and can decompose diastolic murmurs into a series of Intrinsic Mode Functions (IMFs), then accurate IF estimation can be acquired by TEO.

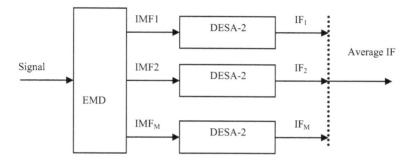

Figure 25. Block diagram of Instantaneous Frequency (IF) estimate based on EMD-TEO

The block diagram of the instantaneous frequency estimate based on EMD-TEO is shown in figure 25 (IF refers to the instantaneous frequency in the block diagram).

The instantaneous frequency of the original signal can be obtained in the following steps:

a.

$$a(n) = 1 + 0.6\cos(0.01\pi n)$$

Decompose the original single into IMFs:
$$\phi(n) = \frac{\pi}{10}n + \cos\frac{\pi}{80}n \qquad j{=}1\ldots M.$$

b. Calculate the instantaneous frequency $c_j(t)$ of each IMF $IF_j(t)$ by DESA-2.

c. Calculate the average instantaneous frequency of the original signal:

$$\omega(t) = \sum_{j=1}^{M} IF_j(t) / M \tag{32}$$

It is the average frequency of mainly IMFs at each instant time.

Next we estimate the IF of diastolic murmurs from clinical coronary artery disease (CAD) patient based on the EMD-Teager method. The left anterior descending artery is stenosed about 40% and the right coronary artery is stenosed about 55%, which has already been diagnosed by catherization. Figure 26 shows the diastolic murmurs. Figure 27 shows the IMFs obtained by EMD. The diastolic murmurs can be decomposed into six IMFs and one residue. The amplitudes of IMF5 and IMF6 are smaller compared with the original signal. So IMF5 and IMF6 are abandoned. Figure 28 shows the IF of each effective IMF by DESA-2. Figure 29 shows the average IF of diastolic murmurs. Then some features such as mean, standard deviation, etc., can be extracted from the average IF. For the normal subject, figure 30 shows the IF of each effective IMF and figure 31 shows average IF of diastolic murmurs.

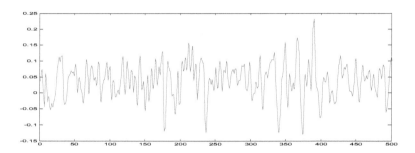

Figure 26. Diastolic Murmurs of CAD object

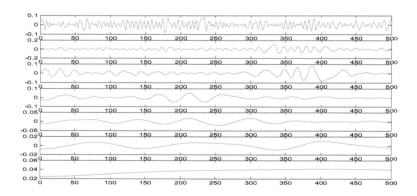

Figure 27. Six IMFs and one residue by EMD

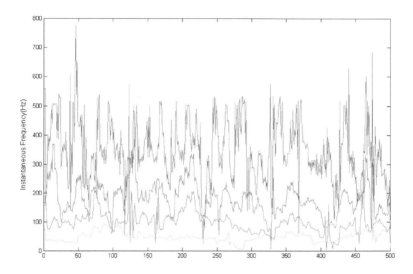

Figure 28. Estimated IF of four selective IMFs by DESA-2

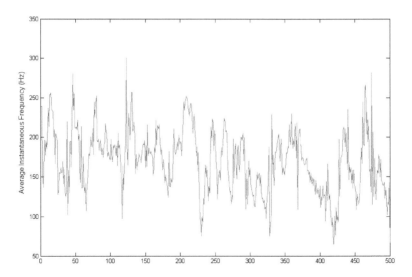

Figure 29. The average instantaneous frequency of diastolic murmurs

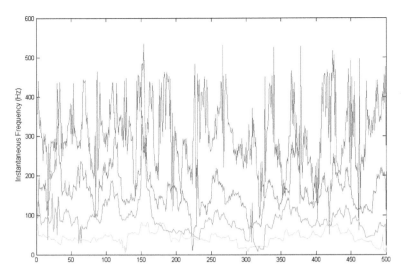

Figure 30. Estimated instantaneous frequency of normal object by DESA-2 algorithm

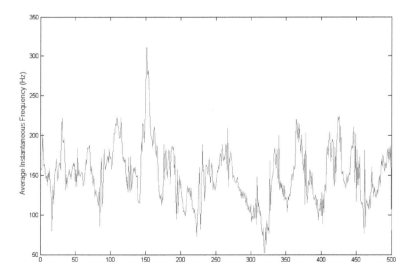

Figure 31. Estimated IF of normal object

For the CAD object, we can see that both the IF of each IMF and average IF are higher than those for normal subject. The diastolic murmurs contain rich higher frequencies. The mean of average IF is 185Hz and the standard deviation is 40Hz. For the normal subject, the mean of average IF is 140Hz and the standard deviation is 26Hz. These can be explained as follows: for the CAD subject, the narrowed coronary arteries lead to the blood flow in coronary artery changing from laminar flow to turbulence flow, from simplicity to complexity. Coronary arterial stenosis gives rise to high frequencies of diastolic murmurs. The instantaneous frequency features effectively reveal the information as to whether the arteries are blocked and denote the frequency change of diastolic murmurs when the coronary arteries are occluded.

6. Conclusion

Diastolic murmurs contain the information of coronary artery occlusions which give the basis of CAD diagnosis. The Hilbert Huang Transform is an adaptive powerful method to analyse nonlinear and non-stationary time series. The important part of HHT is the Empirical Mode composition (EMD). In this paper, we firstly studied wavelet shrinkage denoising using the generalized threshold function and the data-driven SURE threshold value, which successfully removed noise from the PCG signal. Secondly, we obtained the Hilbert spectrum and marginal spectrum of diastolic murmurs for normal subjects and CAD patients after EMD. They provide higher resolution and energy concentration in the time-frequency plane. The Hilbert spectrum and marginal spectrum effectively reveal the information as to whether the arteries are blocked and provide a reliable indicator of CAD. For restricting the end effect of EMD, a simple, powerful and effective method is presented. The IF estimation algorithm is studied based on EMD-TEO. The results indicate that the IF of diastolic murmurs effectively reveal the information on whether the arteries are blocked and provide a reliable indicator of CAD and provides a reliable indicator of coronary artery disease.

Acknowledgements

This paper is partly supported by the National Natural Science Foundation of China (grant no. 61102133) and supported by the Key Project of the Science Technology Department of Zhejiang Province (grant no.2010C11065) and the project of Hangzhou Science and Technology Committee (grant no. 20110833B31).

Author details

Zhidong Zhao[1*], Yi Luo[1], Fangqin Ren[1], Li Zhang[1] and Changchun Shi[2]

*Address all correspondence to: mailzzd@yahoo.com.cn

1 Hangzhou Dianzi University, Hangzhou, China

2 Hangzhou Normal University, Hangzhou, China

References

[1] Akay, M, et al. Harmonic decomposition of diastolic heart sounds associated with coronary artery disease. Signal Processing (1995). , 41(1), 79-90.

[2] Akay, M, et al. Comparative study of advanced signal processing techniques for detection Coronary Artery Disease. Proceedings of the Annual International Conference of the IEEE Engineering in Medicine and Biology Society (1991). , 2139-2140.

[3] Djebbari, A. Bereksi Reguig F. Short-time Fourier transform analysis of the phonocardiogram signal. The 7th IEEE International Conference on Electronics, Circuits and Systems (2002). , 844-847.

[4] Debbal, S. M. Bereksi Reguig F. Time-frequency analysis of the first and the second heartbeat sounds, Applied Mathematics and Computation (2007). , 184(2), 1041-1052.

[5] Khadr, L, Matalgah, M, et al. The wavelet transform and its applications to phonocardiogram signal analysis, Medical informatics (1991). , 16(3), 221-227.

[6] Huang, N. E, et al. The empirical mode composition and the Hilbert spectrum for non linear and non-stationary time series analysis, Proceedings of the Royal Society of London An(1998).

[7] Cheng, J, et al. Research on the intrinsic mode function (IMF) criterion in EMD method, Mechanical Systems and Signal Processing (2006). , 20(4), 817-824.

[8] Gao, H. Y, & Bruce, A. G. Understanding waveshrink: variance and bias estimation, Biometrika (1996). , 83(4), 727-745.

[9] Gao, H. Y. Wavelet shrinkage denoising using the non-negative garrote, Journal of Computational and Graphical Statistics (1998). , 7(4), 469-488.

[10] Stein, C. Estimation of the mean of a multivariate normal distribution, Annuals of statics (1981). , 9(6), 1135-1151.

[11] Zhao, Z, & Wang, Y. A new method for processing end effect in Empirical Mode Decomposition, Communications,International Conference on Communications, Circuits, and Systems (2007). , 841-845.

[12] Gauthier, D, et al. Spectral Analysis of Heart Sounds Associated With Coronary Occlusions 6th International Special Topic Conference on Information Technology Applications in Biomedicine (2007). , 49-52.

[13] Oliveira, P. M, & Barroso, V. Definitions of Instantaneous Frequency under physical constraints. Journal of the Franklin Institute (2000).

[14] Maragos, P, et al. On separating amplitude from frequency modulations using energy operators, ICASSP (1992).

[15] Zhao, Z. D, Zhao, Z. J, et al. Time-frequency analysis of heart sound based on HHT. International Conference on Communications, Circuits and Systems (2005).

[16] Zhao, Z. D, & Pan, M. Instantaneous Frequency Estimation of Diastolic Murmurs Based on EMD and TEO. 1st International Conference on Bioinformatics and Biomedical Engineering, (2007). , 829-832.

[17] Zhao, Z. D. Wavelet shrinkage denoising by generalized threshold function, International Conference on Machine Learning and Cybernetics (2005). , 5501-5506.

[18] Yoshida, H, Shino, H, & Yana, K. Instantaneous frequency analysis of systolic murmur for phonocardiogram,19th Annual International Conference of the IEEE Engineering in Medicine and Biology Society (1997).

Chapter 5

Performance of Adaptive Hybrid System in Two Scenarios: Echo Phone and Acoustic Noise Reduction

Edgar Omar Lopez-Caudana and
Hector Manuel Perez-Meana

Additional information is available at the end of the chapter

1. Introduction

The adaptive noise cancellation has proved being very efficient method in various practical applications such as voice clearance, recognition systems for voice, hands-free telephony, and medical applications such as hearing aids and fetal electrocardiography [1], etc. Figure 1 [1], depicts the basic principle of noise cancellation (understanding that noise is an unwanted signal, d(n)), which is described by main signals that feed the system.

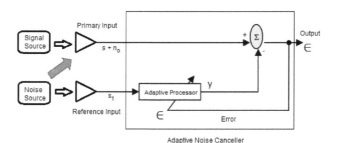

Figure 1. Adaptive noise cancelling approach

Acoustic noise has been studied in recent years due to growing interest in cancelling acoustic noise through active control, since it is increasingly common to find sources of noise in many industrial processes. Basic outlines of noise cancellation were based on the application of passive attenuators that were used for many years without much success [2], however, development of digital signal processing has become increasingly feasible systems active noise cancellation. Active noise cancellation systems cancel unwanted acoustic noise based on the superposition principle: an acoustic noise of equal amplitude but opposite phase is generated in order to cancel out the unwanted noise.

This work discusses a scheme of active noise cancellation using adaptive algorithms of the digital filters required for the correct operation of the proposed system. The signal generation "anti-noise" to cancel the primary source of noise is a problem different from change of environment, since the signal is generated by electrical means and must be propagated acoustically to have the desired effect; this creates a delayed signal in the generation and propagation, so this change is necessary to calculate the required signal. This work considers the estimation of this modification done "offline" [2].

Hybrid ANC systems correspond to a combination of control structures from the feedback and feedforward systems, where the cancelling signal is generated based on the outputs of both the reference sensor and the error sensor. While the feedforward system attenuates the primary noise, which is correlated with the reference signal, the feedback system cancels the predictable components of the primary noise signal that are not observed by the reference sensor.

As an example of the efficiency of the adaptive hybrid systems, this work evaluates a Hybrid Active Noise Control (HANC) system under feedback acoustic situation. Proposed scheme objective is to compare the performance of HANC versus common references: feedback, feedforward and neutralization systems; the inner nature of HANC gives two main characteristics: on line modeling of secondary path and a good performance under acoustic feedback conditions. In the evaluated system, two least mean square (LMS) adaptive filters are used in the noise control process: one for the feedforward stage and the other for the feedback stage; both of them use the same error signal as used in the adaptation of the modeling filter. Then, the combination of the feedback and feedforward stages, results in a solid robustness for the system in acoustic feedback situation.

This chapter discusses a vital application in telecommunications processes, which is the echo in telephone line and the same time a new proposal: the hybrid structure proposed as a solution to this problem. Finally, the computer simulations are presented to show the success of the proposed system. So, this chapter presents an adaptive hybrid system to resolve the problems described: the noise cancellation using adaptive filtering and one proposal for echo cancellation system. Furthermore, we present a hybrid structure which consists of a feedforward structure, used to estimate the noise path, and a feedback structure, used to cancel the noise, i.e., the unwanted signal: echo in telephony systems or noise signals like conversations, snoring or engines. Hybrid active noise cancellation systems are a good solution to these two important problems, since they have the properties of both the feedforward and feedback systems.

2. Adaptive systems as a solution to problems of signal cancellation

2.1. Adaptive Filtering: Active Noise Control

An adaptive filter responds to changes in its parameters like its resonance frequency, input signal or transfer function that varies with time, for example. This behavior is possible since the adaptive filter coefficients vary over time and are updated automatically by an adaptive algorithm. Therefore, these filters can be used in applications where the input signal is unknown or not necessarily stationary. An adaptive filter is composed of two parts: digital filter and adaptive algorithm.

One of the most important applications for this kind of system is active noise control (ANC). ANC systems must respond to changes in frequency of the primary noise they want to cancel out. In other words the primary non-stationary noise varies; hence we must use some kind of adaptive system, to get an acceptable cancellation that carried out many operations at a high speed. The ability of an adaptive filter to operate and respond satisfactorily to an unknown environment, and variations that may be involved in signal reference, to make a powerful adaptive filter for signal processing and control applications. There are several types of adaptive filters but generally all share the characteristic of working with an input signal (input vector), and a desired response (output vector). These two signals are used to compute an estimate of error (error signal), which allows control of the coefficients of the adjustable filter.

In other words, ANC is an approach to noise reduction and a secondary noise source that destructively interferes with the unwanted noise is introduced. In general, active noise control systems rely on multiple sensors to measure the unwanted noise field and the effect of the cancellation. The noise field is modeled as a stochastic process, and an adaptive algorithm is used to adaptively estimate the parameters of the process. Thus, active noise control involves an electroacoustic or electromechanical system that cancels the primary (unwanted) noise based on the principle of superposition; specifically, an anti-noise of equal amplitude and opposite phase is generated and combined with the primary noise, thus resulting in the cancellation of both noises. ANC is developing rapidly because it permits improvements in noise control, often with potential benefits in size, weight, volume, and cost. Thus, the active noise control has been object of an intense research and central subject in many scientific articles in the last 10 years.

On the other hand, unwanted acoustic noise is a by-product of many industrial processes and systems. This problem has become more and more evident as the applications of electronic communication systems increase, since their effects represent an important source of annoyances for the end user and they can reduce considerable the efficiency, the quality and the reliability of this type of systems. These ANC systems use an active form of noise control which includes the use of a second source of sound that generates a signal of the same characteristic as echo but with different phase. This allows to cancel this signal because the waves of sounds propagate linearly, which is known as superposition effect Also, since the characteristics of the signal to cancel change constantly, in this case the echo, the system requires a great capacity of adaptation. These adaptively systems, represent a feasible alternative for echo cancellation in telephone lines due to their processing, capacity and lower cost.

2.2. Cancelling Telephone Echo

Telephone echo, is a phenomenon produced by the mismatching impedance of the hybrid circuit used to couple the two lines with the four lines sections of long distance communication systems that considerably degrades the quality of telecommunication systems. Several systems have been proposed in the literature during the last several years, to solve these problems, such as adaptive echo cancelers. The figure 2 depicts the basic structure of described system.

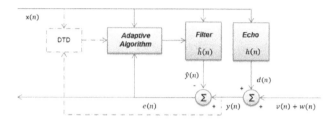

Figure 2. Echocancelling in long distance telephone systems

An echo canceler generates a replica of the echo signal and subtracts it from the signal to be transmitted generating the so-called pseudo echo, which is then used to update the echo canceler coefficients such that the mean square value of residual echo becomes a minimum. However the real time estimation of the hybrid impulse response is a difficult task for several reasons:

1. The echo path impulse response is non-stationary, and then the convergence of adaptation algorithm must be fast enough to track these changes.

2. The power spectral density of speech signals is not flat. This fact results in a slower convergence rate when gradient search based adaptive algorithms are used.

3. In most cases the echo canceler requires one hundred or more taps for an accurate estimation of a hybrid impulse response and several thousand of taps in the acoustic echo path case, which makes the use of efficient adaptation algorithms difficult.

4. The presence of both, near and far-end speakers simultaneously often occurs, which require some robust mechanisms or adaptation algorithms to handle it. Thus the development of low complexity and high convergence rate echo canceler structures has received a lot of attention, resulting in several efficient echo canceler structures and adaptation algorithms.

The most suitable tool for solving the two aforementioned problems is adaptive filtering which has been successfully applied in the solution of several practical problems [3].

3. ANC Systems: types and problematic

3.1. Types of ANC Systems

3.1.1. A priori (Feedforward)

Figure 4 shows, in a simplified way, an ANC Feedforward System, in which the digital filter W(z) is used to estimate the unknown plant P(z). It is assumed that both the plant and the filter have the same input signal x(n). Moreover, a Filtered LMS (Filtered-X Least Mean Square, FXLMS) algorithm is introduced, which is a varying form of the LMS algorithm [2]. FXLMS algorithm solves the secondary path problem, described as the set of transformations that the filter signal and the adaptive error signal go through, on their way from an electric to an acoustic domain. During this electro-acoustic process, the signal may be delayed or altered in such a way that it is necessary to minimize such effects. The FXLMS algorithm technique consists of placing a filter, with the same properties as the secondary path, in the reference signal going towards the adaptive least mean square filter (LMS), as shown in figure 3.

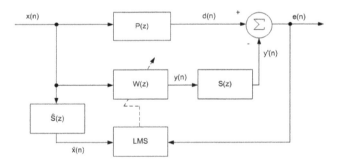

Figure 3. ANC Feedforward system with FXLMS algorithm

From Figure 3, filter $\hat{S}(z)$ is the model of the secondary path, defined by filter S(z). Taking this into consideration, the update of filter W(z) is given as follows:

$$\bar{w}(n+1) = \bar{w}(n) + \mu \hat{x}(n)e(n) \tag{1}$$

Where

$$\hat{x}(n) = \hat{s}(n) * \bar{x}(n)$$

(2)

3.1.2. A posteriori (Feedback)

There are some situations in which it is not possible to take into account the reference signal from the primary noise source in a Feedforward ANC system, due to difficult access to the source, or another reason that makes it hard to identify a specific signal through the reference microphone. A solution to this problem is to introduce a system, which will predict the behavior of the input signal; this system is known as *a posteriori* ANC (Feedback ANC), which is known for using only an error sensor and a secondary sound source to achieve noise control.

Figure 4 describes a Feedback ANC system with FXLMS algorithm, in which d(n) is the noise signal, e(n) is the error signal, defined as the difference between d(n) and signal y'(n), which is the adaptive filter's output once the secondary path has been crossed. Finally, the adaptive filter's input signal is generated by the sum of the error signal and the resulting signal from the convolution between the secondary path $\hat{S}(z)$ and the estimated output of the adaptive filter, y(n).

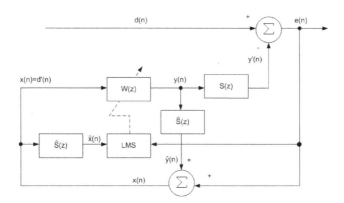

Figure 4. Feedback ANC system with FXLMS algorithm

3.1.3 What is a Hybrid system?

A hybrid ANC system is made up of an identification stage (*feedforward*) and a prediction stage (*feedback*). The combination of both stages needs two reference sensors: one close to the primary noise source and other with the residual error signal. Figure 6 shows the detailed block diagram of a hybrid ANC system, in which it is possible to observe the basic systems (*Feedforward*, *Feedback*) involved in the design. The attenuation signal, given by y(n), results from the addition of both adaptive filter outputs, W(z) and M(z). Filter M(z) represents the *Feedback* process of the adaptive filter, while filter W(z) represents the *Feedforward* process.

The secondary path in the basic ANC system is also taken into consideration in the hybrid system, and is given by the transfer function S(z).

Among the advantages of hybrid ANC systems we can mention:

1.   The fact that lower order filters may be used to achieve the same performance;

2.   The other two systems present much more significant plant noise than the hybrid system;

3.   The combination of both systems allows for much more flexibility in regards of design; and,

4.   Cancellation of both narrowband and broadband noise.

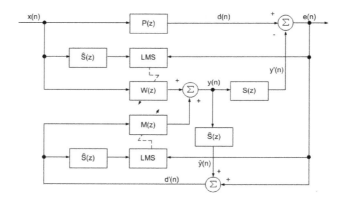

Figure 5. Hybrid ANC system with FXLMS Algorithm

The block diagram if the hybrid ANC system in Figure 5 also shows FXLMS algorithm to make up for the possible delays or problems induced by the secondary path [4].

3.2. Main problems in ANC Systems

3.2.1. Secondary Path Modeling

As mentioned previously, the process that transforms the resulting signal from the adaptive filter y(n) into signal e(n), is defined as secondary path. This characteristic takes into consideration the digital to analog converter, the reconstruction filter, the sound source, the amplifier, the acoustic path from the sound source to the error sensor, the error microphone, and the analog to digital converter. There are two techniques to estimate the secondary path, both with characteristics that make each method more comprehensive and sophisticated in certain ways; these techniques are: offline secondary path modeling and online secondary path modeling. The first method is performed with a *Feedforward* system, where the plant is now S(z) and the coefficients from the adaptive filter are the secondary path estimation, as shown in Figure 6 [4].

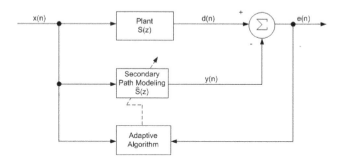

Figure 6. Offline Secondary Path Modeling

3.2.2. Acoustic Feedback

This property is typical of feedforward systems. Figure 7 shows the contribution of attenuation signal y(n), which causes the system to degrade because of the signal present in the reference microphone.

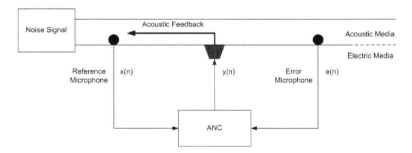

Figure 7. *Feedforward* ANC process with acoustic feedback

Two possible solutions for acoustic feedback problem are: acoustic feedback neutralization and the proposal of a hybrid system, which has a better performance in the frequency range and attenuation level of interest [4]. To evaluate this approach, we used a hybrid system as shown in Figure 8, where F(z) is the transfer function of the feedback process.

The system proposed in [5] will be analyzed and this system, with a set of signals and experimental conditions, was completely evaluated in [6].

3.2.3. Online Acoustic Feedback Path Modeling

Most common way to eliminate acoustic feedback is to make an online path modeling, like indicated on [3] and, more recently, in relevant papers by [7] and [8]. However, one of the main characteristics of the hybrid system presented by the authors in [9] is that it does not take the secondary path modeling into consideration. Instead, the proposed hybrid system takes advantage of the inherent robustness of hybrid systems when it comes to acoustic feedback, figure 8.

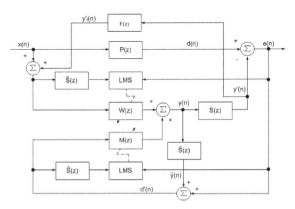

Figure 8. Hybrid ANC system with acoustic feedback

The system in Figure 9, proposed by [10], was used to compare the robustness of the HANC system against the neutralization system.

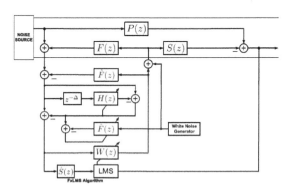

Figure 9. Kuo's Neutralization System

The details of the system in Figure 9 can be consulted in [10]. However, an important fact of this system is that it uses additive noise for modeling, also, as mentioned in [7], regarding predictable noise sources.

4. Echo Cancellation

4.1. Definition and general review

The echo is a problem that significantly degrades the quality of telecommunication systems. This occurs, in telephone line, due to the decoupling impedance hybrid which exists in the coils and are used to couple subscriber communication channel with the long distance channels. There is also the so-called acoustic echo which occurs in teleconferencing systems and hands free telephone systems. This type echo occurs due to acoustic coupling between loud-speakers and microphones used in these communication systems.

Several systems which try to solve this problem have appeared in the literature in recent years. Among these are: directional microphone arrangement [11], echo suppressors and adaptive echo cancellers [11, 12], etc. Among them, adaptive echo cancellation seems to be the best way to reduce the echo problem [13, 14]. An echo canceller generates an echo replica and subtracts the signal to be transmitted, generating a so-called residual echo. The echo residual is then used to adapt the coefficients of the system, using in most cases a gradient-based algorithm, in a way that the mean square value of the residual echo is progressively minimized [11, 12, 13, 14]. However, the real-time estimate of the impulse response of the hybrid or echo channel is a complex problem for several reasons:

1. The duration of the impulse response of a typical echo channels in teleconferencing systems in the order of several hundreds of milliseconds, which means that transversal filter coefficients of several thousand would be needed to reduce echo to acceptable levels. The impulse response of a typical acoustic echo channel is shown in Figure 10.

2. The impulse response of echo channel is non-stationary because it changes with the movement of the interlocutors, or the number of active subscribers on a given time. Thus the adaptive algorithm should be fast enough to track those changes.

3. The power density spectrum of the voice is not flat, and in many cases reduces speed of convergence of the adaptive algorithm. The correct estimate of the echo channel using structures with the least possible complexity and the relatively high speeds obtain convergence of the adaptation algorithm, as mentioned above, are non-trivial problems which have received considerable attention in recent years; among the different proposed have been proposed several echo cancellation systems, among we can mention: transverse echo cancellers, echo cancellers in the frequency domain, echo cancellers infinite impulse response subband echo cancellers, etc., [11, 12, 13, 14].

Besides the reduction in the complexity of the canceller, to allow correct estimation of the echo channel and the development of adaptive algorithms with rapid convergence, another major problem is handle the simultaneous presence of echo near the speaker's voice. The situation we want to avoid is to interpret the speaker's voice echoing nearby, and make great changes in the echo channel estimated in an unsuccessful attempt to cancel this. A checked algorithm could operate incorrectly when the distant partner is present, so it is necessary to incorporate certain mechanisms within the system to avoid this effect [11, 12, 13, 14].

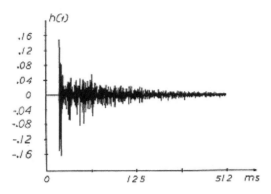

Figure 10. A typical impulse response of acoustic echo channel

There are few references about the convenience of using adaptive hybrid schemes for solving the problem of echo cancellation, and given the results obtained for applications for cancellation of acoustic noise [15], hybrid scheme is proposed for electrical noise cancellation, since it is on the phone lines where there is the problem described. Be detailed later about how to do and the results achieved.

4.2. Telephone Systems

A long distance telephone system basically consists of a 2-wire portion, known as the subscriber circuit, and connects the subscriber to the local exchange and long-distance circuits itself; this system consists of a transmission channel and another receiving, each of which consists of two wires. A hybrid transformer is used to couple circuits' long distance subscriber circuit and ideally isolate the transmission channels and reception of long-distance circuit. However due to the decoupling impedance, they are not completely isolated so that a portion of the received signal is delayed in the form of echoes. A similar problem arises in teleconferencing systems with so-called acoustic echo which occurs due to coupling between the microphone and speaker in the teleconference system. This result in a delayed and distorted replica of the signal produced by the loudspeaker is fed back into the microphone.

In both cases there is deterioration in the communication system, which resulted in the appearance of echo cancellers. These cancellers have proved to be the best way to solve this problem [11, 12]. The basic principle of echo cancellation, which is illustrated in Figure 11, is to generate an echo replica, this is subtracted from the signal to be transmitted, resulting in the so-called residual echo that consists of part of the signal echo which could not be canceled more near the speaker's voice, if this is present [11, 12]. The residual echo is then used to adapt the parameters of the canceller in such a way that the residual echo power is progressively minimized.

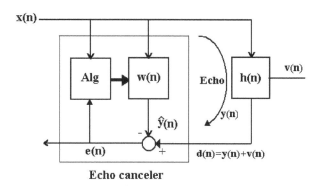

Echo canceler

Figure 11. Echocancelling in long distance telephone systems

Echo canceller consists of two main parts. An adaptive filter, which generates an echo replica and is subtracted from the signal being transmitted, and a system commonly known as double-talk detector, this system prevents distortion due to the presence of the speaker's voice service or in the absence of the partner away. The first component is the structure of the adaptive filter along with its adaptation algorithm.

Some researchers have resulted in the appearance of various structures, such as transversal filters, subband structures, structures in the frequency domain, etc., and various adaptive algorithms, mostly based on gradient descent search. Second component, despite its importance, has received much less attention than the first component. Thus conducting research aimed at developing highly reliable mechanisms to avoid distortion due to the simultaneous presence of both parties, "double-talk detector," especially when using algorithms based on gradient descent search is of great importance.

5. Delimitation of Proposed ANC System and its Application

5.1. Evaluated ANC Structure

Figure 12 shows the block diagram of the evaluated hybrid ANC structure with online secondary path modeling. This hybrid ANC structure consists of a feedforward stage, W(z), which is used to estimate the noise path, P(z), and a predictive structure, M(z), which is used to cancel the distortion due to the acoustic feedback path, F(z). Since the samples of feedback distortion are strongly correlated among them, they can be predicted [15].

As shown in Figure 12 signal, $a(n)$, is used simultaneously as:

1. The error signal to update the adaptive filter, $W(z)$, which corresponds to the feedforward stage used to identify the noise path, and,

2. To update the linear predictive filter $M(z)$, which intends to cancel the distortion produced by the feedback propagation from the canceling loudspeaker to the input microphone thorough the system $F(z)$; and,

3.
   To estimate $\hat{S}(z)$, which represents the online secondary path modeling adaptive filter.

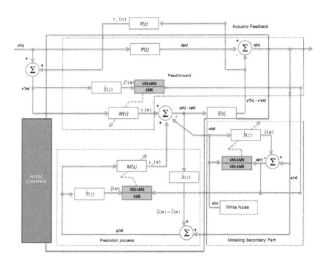

Figure 12. Evaluated hybrid ANC structure

The hybrid ANC contains the advantages of feedback and feedforward systems. The model presented by [16] was modified to adapt the system for a specific objective: reduce the residual echo. This system uses two input signal $x(n)$ and $din(n)$, one for each talker. The plant that models echo refers to the effect of mismatch of impedance present in the telephone circuit. The echo signal is $d(n)$ and the residual echo plus the far-end signal is represented by $e(n)$. This system incorporates the signal of the feedforward and the feedback effect that means both systems contribute to generate the cancelling signal, which approximates to the echo signal. Also this system includes a switch on the feedback system: when the echo signal and the far-end signal are highly correlated, the feedback system cancels part of the far-end signal even if the hybrid system already converged [17].

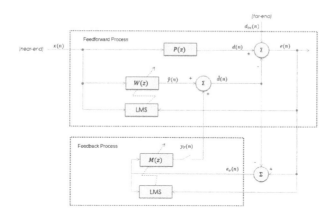

Figure 13. Adapted Hybrid ANC for Active Echo Cancellation

To analyze the system is necessary to consider the correlation between signals, as shown in the equation (3):

$$R = E\left[\bar{x}(n)\bar{x}^2(n)\right] \tag{3}$$

The cross correlation vector between the entrance and the echo is given by:

$$\bar{p} = E\left[d(n)x(n)\right] \tag{4}$$

and the correlation matrix can be written as follows:

$$R\overline{w_0} = \bar{p} \tag{5}$$

where $\overline{w_0}$ is the optimum vector of the transversal filter. In the selected algorithm, LMS, the reference signal $x(n)$ is processed by an adaptive filter $W(z)$. In this case the coefficients of the filter are updated by the gradient of the error signal power obtained plus the previous coefficients and μstep size:

$$w(n+1) = w(n) + \mu x(n)e(n) \tag{6}$$

5.2. Active Echo Cancellation in Telephone Lines

There are two kinds of echo: electric and acoustic. The electric echo is present in traditional telephony lines because of the impedance mismatch of the conversion (from two to four wires). The acoustic echo is the direct or indirect feedback of reflected signals to the micro-phone during a conversation. There are two controls applied to echo: suppressor and cancel-ler systems. Echo Cancellation systems need to consider the disturbances in the far-end talker's signal and the superposition of the near-end talker's that generates double-talk. Two general approaches are the use of suppressors and the use of cancellers. The echo suppres-sor has a sensor that measures the voice signal power in each part of the circuit to decrease the impact of the echo. The echo suppressor changes the full duplex channel to a half-duplex channel [14, 18]. This characteristic is a disadvantage of this type of control because it can-cels part of the speech. Echo cancellers use the superposition principle that means this sys-tem generates a similar signal with delay and attenuation similar to the transmitted signal. It is recommended to train the system to approach the characteristics of the echo signal. For this problem some authors [19, 20], offered different solutions based on Double-Talk Detec-tor (DTD) [21]; this principle detects the presence of simultaneous speech of both talkers and pause the coefficient updating of the adaptive filter. It is known that the adaptive filter is the key to treat echo problems. It is necessary to consider the speed of convergence and robust-ness of the system. Most of echo cancellation systems use transversal filters and the LMS al-gorithm or variations of this to adjust the coefficients [22].

The result is an error signal named as residual echo signal due to estimation of the adaptive filter [21], this scenario, adapted to an ANC system is shown in Figure 14 [3].

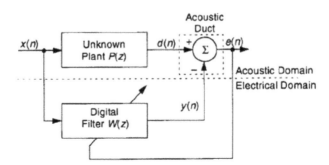

Figure 14. System identification viewpoint of ANC

From Figure 14, the residual echo e(n) is defined as

$$x(n) = d(n) - y(n) \tag{7}$$

where $d(n)$ is the echo signal and $y(n)$ is the response generated by the adaptive filter after processing the algorithm. Also [3], presents the criteria of the Mean Square Error (MSE) to find the convergence point of the system. To analyze the performance of the Echo Cancellation system Echo Return Loss Enhancement (ERLE) criteria was developed. The ERLE criterion is described in equation (8).

$$ERLE = 10\log\left\{\frac{E\left[d^2(n)\right]}{E\left[e^2(n)\right]}\right\}$$ (8)

ERLE parameter was used to evaluate the present proposed system.

6. Performance Parameters and Several Aspects Considered

6.1. Parameters and issues

The proposed system has different parameters to consider. These parameters determine whether the system converges or not.

1. *Step size (μ):* controls the system stability and speed of convergence, one for each part of the system (feedback and feedforward).

2. *Plant:* simulates the echo effect

3. *Adaptive filter $W(z)$:* length and values for established plants

4. *Number of blocks and iterations:* reflected in the number of samples observed

5. *Entrance signals:* including the near-end and the far-end

Step size values were taken by [16, 23]. The plant simulates the effect of echo that the near-end suffer because of the impedance mismatch, proposed by [24].

The input signals utilized are sorted into one of three types, considering the classification proposed by [3, 25], as well as companies such as [26].

1. *Continuous;* the level of sound remains constant or nearly constant with small fluctuations. For Echo cancellation, the selected signals were vacuum, four tones and silence.

2. *Intermittent:* the level of sound presents some fluctuations that can be periodic or random. The selected signals are real voices recorded in a computer for Echo considerations.

3. *Impulsive:* the level of noise presents impulses in a brief period of time.

For Acoustic Noise Reduction applications, the system was tested with several real sound signals taken from an Internet database [27]. The sound files were selected taking into account that the system is to be implemented in a duct-like environment. Also, six different types of signals were used for the analyzed system:

1. A sinusoidal reference signal with frequency of 300 Hz, and 30 dB SNR;

2. A reference signal composed of the sum of narrow band sinusoidal signals of 100, 200, 400, and 600 Hz; and,

3. The rest of the reference signals are.wav audio files with recordings of real noise sources, which are *"motor"* and *"airplane"*, as in [16].

The most important values are modeling error, as was defined by [28], and MSE, given by the ratio between the power of the error signal, and the power of the reference signal:

$$\Delta S(dB) = 10 \log_{10} \left[\frac{\sum_{i=0}^{M-1} [s_i(n) - \hat{s}_i(n)]^2}{\sum_{i=0}^{M-1} [s_i(n)]^2} \right] \tag{9}$$

$$MSE(dB) = 10 \log_{10} \left[\frac{\sum_{i=0}^{M-1} [e_i(n)]^2}{\sum_{i=0}^{M-1} [x_i(n)]^2} \right] \tag{10}$$

6.2. System Training

For experience, we need to train the system before to start to work [16]. So, we have two considerations:

1. For echo cancellation, we adapt the plant for 20 representative coefficients instead the 1000 given by [24]. The adaptive filter was a vector of 20 coefficients initialized in zero. The near-end voice was a female voice and silence for the far-end. The step size value were change until get the higher level of ERLE, after run the simulation of the system using Matlab®, with a software interface developed specifically for this purpose, the results of the adaptive filter were retaken to repeat the processing, when a 40dB of cancellation were achieve the training was stopped. The scenario for training work was single-talk with a single voice signal in the near-end.

2. For the situation for Acoustic Noise Reduction, secondary path is offline modeling stopped when the error is reduced-35dB similar to [15]. The excitation signal v(n) used was white Gaussian noise with variance of 0.05.

7. Analysis of Results

7.1. Echo cancellation phone lines

To consider an approximation of a real system the results of processing echo of voice with the hybrid proposed system. We present the results using the female voice signal (Figure 15) in the near-end and two different masculine voice signals in far-end (Figure 16 and Figure 17).

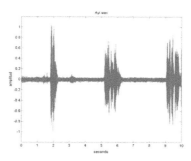

Figure 15. Female voice signal

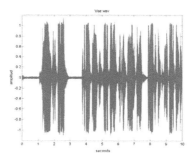

Figure 16. First masculine voice signal

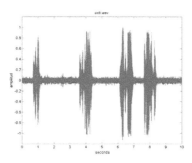

Figure 17. Second masculine voice signal

The echo signal generated by the adaptation of the plant is represented in the Fig 18.

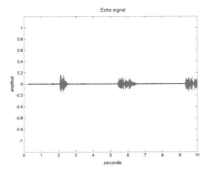

Figure 18. Echo of the female voice signal with adapted plant

Applying the function with the parameters of Table 1, the obtained results are shown in Figure 19 and Figure 20. Both figures show that system achieves cancellation of the echo signal.

Parameters	Value
Step size	0.1
Plant	From [24]
Blocks	1000
Iteration	80

Table 1. Analysis Parameters

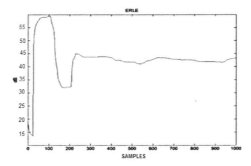

Figure 19. ERLE using female voice in the near-end and masculine voice 1 in far-end

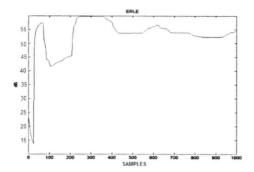

Figure 20. ERLE using female voice in the near-end and masculine voice 2 in far-end

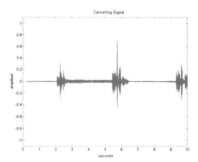

Figure 21. Cancelling voice signal, system with masculine voice 1

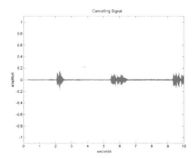

Figure 22. Cancelling voice signal, system with masculine voice 2

Looking for a detailed analysis in the cancelling signal (Figure 21), which imitates echo signal, for the first masculine signal, the system begins to diverge. This occurs because of the high correlation between the two entrances voices; this effect is given by the feedback because even when the system already converge starts to cancel the far-end signal [29].

Then, instead of the first male signal, another signal was used and the system converged better, this can be seen in Figure 22, this situation is because the correlation between this signal and the female is smaller.

As mentioned before, the step size factor has a major impact on the development of the system, and proved to be the main reason to make the system converge; additional simulations were performed using the parameters in Table 2; this means a smaller size step and the male voice first.

Parameters	Value
Step size	0.01
Plant	From [24]
Blocks	1000
Iteration	80

Table 2. Analysis Parameters for Additional Test

The system improves its performance using the parameters of Table 2. The generated cancelling signal (Figure 23), does not have impulsive periods.

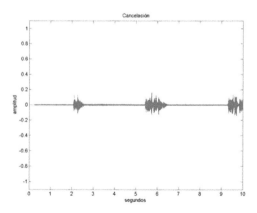

Figure 23. Cancelling voice signal, system with masculine voice 1 and adjusted step size

7.2. Active Noise Cancellation

7.2.1. General results (MSE and Modelling Error)

This section presents the simulation experiments performed for acoustic noise reduction. First, an offline modeling was used to obtain FIR representations of tap weight length 20 for $\hat{P}(z)$ and of tap weight length 20 for $S(z)$. The control filter $W(z)$ and the modeling filter $\hat{S}(z)$ are FIR filters of tap weight length of $L = 20$ both of them. A null vector initializes the control filter $W(z)$. To initialize $\hat{S}(z)$, offline secondary path modeling is performed, which is stopped when the modeling error has been reduced to -5dB. The step size parameters are adjusted by trial and error for fast and stable convergence.

Various articles on the subject of ANC were references taken into consideration before establishing main analysis parameters to determine the hybrid system's performance:

a) Filter order; it is important to evaluate the system under filters of different orders. In this case, 20 coefficients were selected (we considered the fact that the distance between the noise source and the control system is not supposed to be very large).

b) Nature of the filter coefficients; on a first stage, the coefficients were set according to real values taken from a previous study made on a specific air duct [2]. These coefficients were taken from the work done in [16] to determine the values of the primary and secondary path filters for an air duct.

The simulation results are presented according to the following parameters:

1. Mean Square Error (MSE); and

2. Modeling error from online secondary path modeling.

Table 3 shows the values used for the feedforward and feedback step sizes, as well as the range of step sizes used for the secondary path filter. The values were set by trial and error, starting with the values that were determined with the previous test.

Signal	Step size μ_w, μ_m	Step size μ_s
Continuous	0.000001	0.0001 – 0.001
Intermittent	0.000001	0.0001 – 0.001
Impulsive	0.000001	0.0001 – 0.001

Table 3. Filters Step Size Used in Proposed Analysis

Also, a white noise with zero mean and variance equal to 0.05 was used in the system. Since there were not enough resources to implement an abrupt secondary path change (which means there was only one set of values available for the secondary path filter from [15], a gradual change was made, given by the sum of a sinusoidal function to the secondary path coefficients, from iteration 1000 to 1100. The best response was shown by the continuous signal; Figure 24 shows the Modeling error for this case, while Figure 25 shows the MSE.

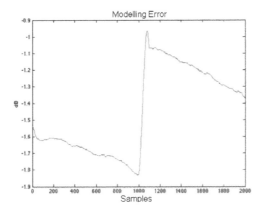

Figure 24. Relative modeling error for continuous signal

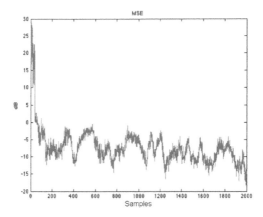

Figure 25. MSE for continuous signal

From Table 3, it can be noticed that the step sizes had to be considerably reduced, in the order of 1000, in comparison to the values established for the tests with Echo cancellation. This is due to the fact that the coefficient values are not necessarily within a range of -1 to 1, so the secondary path modeling needs a smaller step size to be able to achieve convergence.

For the intermittent signal, the effects of the small step sizes were similar: the system took more time to converge and the level of noise cancellation was reduced. Nonetheless, the response achieved stability during the simulation. Figure 26 and Figure 27 correspond to the Modeling error and MSE for the intermittent signal, respectively.

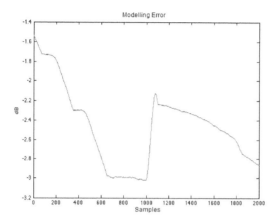

Figure 26. Relative modelling error for intermittent signal

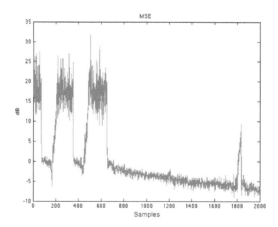

Figure 27. MSE for intermittent signal

Finally, for impulsive input signal the results were not as good as expected. The results can be explained since there are very abrupt changes in the signal amplitude, and the step size is very small. Hence, the values of the coefficients tend to infinity and the simulation stops abruptly.

7.2.2. Comparison versus Neutralization and Feedforward Systems

In this section, three paths were used: the main or primary path P(s), the secondary path S(s), and the acoustic feedback path F(s). All the filters used in the evaluated proposals are finite response filters (FIR). The values of these paths are based on [2], and represent the experimental values of a given duct. A total of 25 coefficients will be used in all paths so as to report an extreme condition for a real duct under analysis.

To initialize $\hat{S}(z)$, the offline secondary path modeling is stopped when the modeling error has been reduced up to -35dB, similar to [15]. The excitation signal $v(n)$, is white Gaussian noise with variance equal to 0.05.

The values for the step size are adjusted by trial-and-error to achieve a faster convergence and stability, following the guidelines from previous work on Hybrid Active Noise Control [16], and the values selected in [7] for neutralization. A summary of the selected values for μ, is shown in Table 4.

System	Primary Path μ_P	Secondary Path μ_S	Feedback Path μ_F
Neutralization System	0.000001	0.00005	0.00005
Hybrid System	0.001	0.001	NA

Table 4. Filters Step Size Used in Proposed Analysis

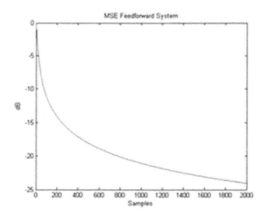

Figure 28. MSE with "*sinusoidal*" reference signal for Feedforward System

Figure 28 to Figure 41 show the result of the systems analysis with the previously mentioned set of signals. All results are shown in dBs, measuring the error power at the output (Mean Square Error).

First, we show the main signal for ANC systems, the sinusoidal signal. Figures 28 to 30 show the MSE value obtained.

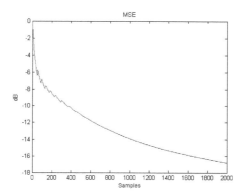

Figure 29. MSE with "*sinusoidal*" reference signal for Neutralization System

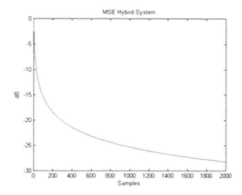

Figure 30. MSE with "*sinusoidal*" reference signal for Hybrid System

Another important is a narrow band signal, as explained before, is composed of the sum of narrow band sinusoidal signals of 100, 200, 400, and 600 Hz. Figures 31 to 33 show the results for this consideration

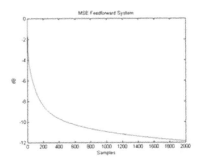

Figure 31. MSE with *"4 tones"* reference signal for Feedforward System

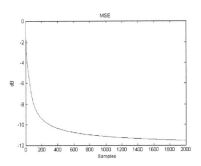

Figure 32. MSE with *"4 tones"* reference signal for Neutralization System

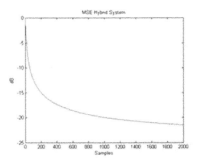

Figure 33. MSE with *"4 tones"* reference signal for Hybrid System

Finally, we use two recorded signals, corresponding to a "plane" and one to a "motor", meaning the evidence most relevant to our system. Of Figures 34 through 41, shows the convergence achieved with the proposed system.

Finally, it is important to consider that an ANC system should respond successfully to a change in the status of secondary path, which corresponds, for example, a possible movement of the microphone in a pipeline, or any vibration or change in of the system. Figures 24 and 25 show an abrupt change in secondary path at iteration 1000 [5]. We can observe that the behavior of both remain stable.

Figures 40 and 41 show selected results for the neutralization and hybrid systems, which are of greatest interest.

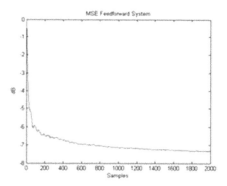

Figure 34. MSE with "*Motor*" reference signal for Feedforward System

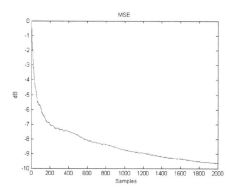

Figure 35. MSE with "*Motor*" reference signal for Neutralization System

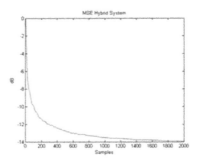

Figure 36. MSE with *"Motor"* reference signal for Hybrid System

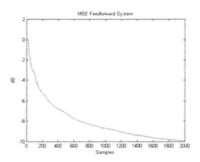

Figure 37. MSE with *"Airplane"* reference signal for Feedforward System

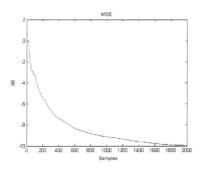

Figure 38. MSE with *"Airplane"* reference signal for Neutralization System

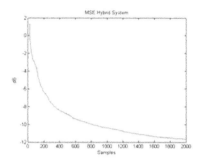

Figure 39. MSE with *"Airplane"* reference signal for Hybrid System

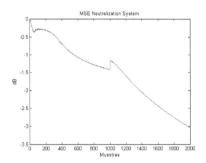

Figure 40. MSE with *"4 tones"* reference signal for Neutralization System, considering changing secondary path

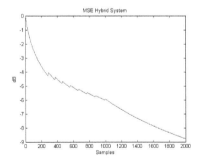

Figure 41. MSE with *"4 tones"* reference signal for Hybrid System, considering changing secondary path

8. Conclusions

The adaptive filtering is a powerful tool that offers various solutions to many fields of science today. This chapter shows the efficiency of the hybrid system in reducing electrical noise and noise currently present in conventional systems where noise becomes a significant cause of health problems, or a situation that can affect communications Internet or phone, to name a few.

Adaptive filtering, which has been successfully applied in the solution of several practical problems which main kinds are described some in this chapter, has relied mainly in the transversal filter structures. However, when the filter order becomes large, the transversal computational complexity and convergence rate may limit its capability for solving practical problems. This chapter presented an overview of the Hybrid System.

In particular, there are few references about hybrid systems, those conjoined feature more traditional patterns such as *a priori* and *a posteriori* systems. Of course they inherit the problems of these two, but the advantage they offer is based on the robustness of such systems for signals of different characteristics as continuous, intermittent and impulsive, and we tested a hybrid system in two interesting and relevant scenarios: unwanted signals in the fields of acoustics and telephony.

The proposed system works in an acceptable way for telephone echo problems, but it is necessary to consider and adjust the different parameters. The system is capable of cancelling echo of voice signals and can be applied to simulated scenarios of double talk without use the Double Talk Detector. Also it is necessary to evaluate the correlation between input signals since this correlation has a great impact of the performance of the system. If both signals are highly correlated, it is necessary to use a small step size for both feedback and feedforward systems. We established the double talk situation in telephony conversations as the test system for our Hybrid system including some talks simulating a real conversation.

With respect to Acoustic Noise Reduction, it must be notice that the results presented for a real-value filter coefficients refer to only one specific type of duct. This means that the response could probably improve in a different environment or in a duct with different properties. This situation represents a problem for the designer of a hybrid ANC, because for each environment where the system is to be applied would be no need to identify accurately the parameters to achieve the desired response. However difficult, this may not be impossible to do, so there is still a lot of work to be done with hybrid ANC systems.

This chapter discusses a new Hybrid Active Noise Control system and the impact adaptive filtering has on this field. The objective is to achieve improved performance at a reasonable computational cost in a Hybrid ANC system that considers two of the more important troubles of the ANC. We show two examples to prove the contribution of this system, one is a little generalist about cancelling several kinds of noise, and one very specific, which represents one persistent problem like telephone echo on telecommunications nowadays: networks have been modified by the use of new technologies and constant innovations have led to automate the process of interconnection of subscribers, and the inclusion of forms of streaming media.

Therefore, he was a rigorous analysis of the results and their parameters under the above considerations. The results show the relevance of hybrid systems for consideration in removing acoustic noise or echo in telephony, with tools of adaptive systems. The advisability of this hybrid system is a matter that must be analyzed in depth.

Acknowledgements

This work was supported by the Department of Mechatronics, part of the School of Design, Architecture and Engineering, Tecnológico de Monterrey, Campus Ciudad de Mexico.

Author details

Edgar Omar Lopez-Caudana[1*] and Hector Manuel Perez-Meana[2]

1 Tecnológico de Monterrey, Campus Ciudad de México, Mexico

2 SEPI, ESIME Culhuacan, IPN, Mexico

References

[1] Widrow & Stearns. (1985). Adaptive Signal Processing. *Prentice Hall, Englewood Cliff, NJ.*

[2] Kuo & Morgan. (1996). Active Noise Control Systems: Algorithms and DSP Implementations,. New York: Wiley *Series in Telecommunications and Signal Processing Editors*1996.

[3] Kuo & Morgan. (1999). Active Noise Control Systems: A tutorial review, *Proceedings of the IEEE*, 87(6), 943-973.

[4] Perez-Meana, H., et al. (2007). Active Noise Canceling: Structures and Adaptation Algorithms, *Advances in audio and Speech Signal Processing: Technologies and Applications*, Idea Group Publishing, Hershey, 286-308.

[5] Lopez-Caudana, E., et al. (2008). A Hybrid Active Noise Canceling Structure, *International Journal of Circuits, Systems and Signal Processing*, 2(2), 340-346.

[6] Lopez-Caudana, E., et al. (2010). Evaluation for a Hybrid Active Noise Control System with Acoustic Feedback, *53rd IEEE Int'l Midwest Symposium on Circuits & Systems*, 1-4.

[7]   Akhtar, E., et al. (2007). Acoustic feedback neutralization in active noise control sys-
      tems, *IEICE Electronics Express*, 4(7), 221-226.

[8]   Akhtar, D. (2007). On active Noise Control Systems with Online Acoustic Feedback
      Path Modeling, *IEEE. Transactions on Audio, Speech, and Language Processing* Febru-
      ary., 15(2), 593-599.

[9]   Lopez-Caudana, E., et al. (2008). A Hybrid Noise Cancelling Algorithm with Secon-
      dary Path Estimation,, *WSEAS TRANSACTIONS on SIGNAL PROCESSING*, 4(12).

[10] Kuo, Sen. M. (2002). Active Noise Control System and Method for On-Line Feedback
      Path Modeling,, *US Patent 6,418,227*.

[11] Pérez-Meana, H., et al. (1994). Echo Cancellation in Audio Terminals, *Memoria Técni-
      ca, MEXICON 94*, 159-164.

[12] Pérez-Meana, H., & Nakano, M. (1990). Cancelación de Eco en Sistemas de Teleco-
      municación, *Mundo Electrónico*, 207, 143-150.

[13] Gritton, C. W., & Li, A. W. (1984). Echo Cancellation Algorithms, *IEEE ASSP Maga-
      zine*, 30-37.

[14] Murano, K., & Amano, F. (1993). Echo Cancelling Algorithms, *Enciclopedia de Teleco-
      municaciones*, 6, Marcel Decker Inc, 383-409.

[15] Lopez-Caudana, E., et al. (2008). A hybrid active noise cancelling with secondary
      path modeling, *Circuits and Systems, 2008. MWSCAS 2008. 51st Midwest Symposium
      on.*, 277-280.

[16] Lopez-Caudana, E., et al. (2009). Evaluation of a Hybrid ANC System with Acoustic
      Feedback and Online Secondary Path Modeling, *19th International Conference on Elec-
      tronics, Communications and Computers 2009, Cholula, Puebla*, 26-28.

[17] Mehmood & Tufail. (2009). A new variable step size method for online feedback path
      modeling in active noise control systems, *Multitopic Conference INMIC 2009. IEEE
      13th International*, 1-6.

[18] Lee, E., & Messerschmitt, D. (1993). *Digital Communication*, Kluwer Academic Pub-
      lisher, Norwell, MA.

[19] Buchner, H., et al. (2006). Robust extended multidelay filter and double-talk detector
      for acoustic echo cancellation, *Audio, Speech, and Language Processing, IEEE Transac-
      tions on*, 14(5), 1633-1644.

[20] Kun & Xiaoli. (2008). A double-talk detector based on generalized mutual informa-
      tion for stereophonic acoustic echo cancellation systems with nonlinearity, *Signals,
      Systems and Computers, 2008 42nd Asilomar Conference*, 2161-2164.

[21] Jae & Dong. (2005). Network echo canceller based on the practical adaptive filter, *In-
      telligent Signal Processing and Communication Systems, 2005. ISPACS 2005. Proceedings
      of 2005 International Symposium on*, 693-696.

[22] Tandon, , et al. (2004). An efficient, low-complexity, normalized LMS algorithm for echo cancellation," Circuits and Systems. *NEWCAS 2004. The 2nd Annual IEEE Northeast Work shop on,* 161-164.

[23] Shoureshi, R. (1994). Active noise control: a marriage of acoustics and control, *American Control Conference,* 3, 3444-3448.

[24] Paleologu, , et al. (2010). An Efficient Proportionate Affine Projection Algorithm for Echo Cancellation *Signal Processing Letters, IEEE* February., 17(2), 165-168.

[25] Romero, A., et al. (2008). A Hybrid Active Noise Canceling Structure, *International Journal of Circuits, Systems and Signal Processing,* 2(2), 340-346.

[26] Brüel & Kjær Sound & Vibration Measurement A/S. Environmental Noise Booklet. (2008). http://www.nonoise.org/library/envnoise/index.htm.

[27] Free Sound Effects, Samples & Music Free Sound Effects Categories. http://www.freesfx.co.uk/soundeffectcats.html (2011).

[28] Akhtar, M. T., et al. (2006). A new variable step size LMS algorithm-based method for improved online secondary path modeling in active noise control systems. *IEEE Transactions on Audio Speech, and Language Processing,* 14(2), 720-726.

[29] Mohammadzaheri, M., et al. (2009). A design approach for feedback-feedforward control systems, *Control and Automation, 2009. ICCA 2009. IEEE International Conference on,* 2266-2271.

Permissions

The contributors of this book come from diverse backgrounds, making this book a truly international effort. This book will bring forth new frontiers with its revolutionizing research information and detailed analysis of the nascent developments around the world.

We would like to thank Dr. Lino Garcia Morales, for lending his expertise to make the book truly unique. He has played a crucial role in the development of this book. Without his invaluable contribution this book wouldn't have been possible. He has made vital efforts to compile up to date information on the varied aspects of this subject to make this book a valuable addition to the collection of many professionals and students.

This book was conceptualized with the vision of imparting up-to-date information and advanced data in this field. To ensure the same, a matchless editorial board was set up. Every individual on the board went through rigorous rounds of assessment to prove their worth. After which they invested a large part of their time researching and compiling the most relevant data for our readers. Conferences and sessions were held from time to time between the editorial board and the contributing authors to present the data in the most comprehensible form. The editorial team has worked tirelessly to provide valuable and valid information to help people across the globe.

Every chapter published in this book has been scrutinized by our experts. Their significance has been extensively debated. The topics covered herein carry significant findings which will fuel the growth of the discipline. They may even be implemented as practical applications or may be referred to as a beginning point for another development. Chapters in this book were first published by InTech; hereby published with permission under the Creative Commons Attribution License or equivalent.

The editorial board has been involved in producing this book since its inception. They have spent rigorous hours researching and exploring the diverse topics which have resulted in the successful publishing of this book. They have passed on their knowledge of decades through this book. To expedite this challenging task, the publisher supported the team at every step. A small team of assistant editors was also appointed to further simplify the editing procedure and attain best results for the readers.

Our editorial team has been hand-picked from every corner of the world. Their multi-ethnicity adds dynamic inputs to the discussions which result in innovative outcomes. These outcomes are then further discussed with the researchers and contributors who give their valuable feedback and opinion regarding the same. The feedback is then collaborated with the researches and they are edited in a comprehensive manner to aid the understanding of the subject.

Apart from the editorial board, the designing team has also invested a significant amount of their time in understanding the subject and creating the most relevant covers. They scrutinized every image to scout for the most suitable representation of the subject and create an appropriate cover for the book.

The publishing team has been involved in this book since its early stages. They were actively engaged in every process, be it collecting the data, connecting with the contributors or procuring relevant information. The team has been an ardent support to the editorial, designing and production team. Their endless efforts to recruit the best for this project, has resulted in the accomplishment of this book. They are a veteran in the field of academics and their pool of knowledge is as vast as their experience in printing. Their expertise and guidance has proved useful at every step. Their uncompromising quality standards have made this book an exceptional effort. Their encouragement from time to time has been an inspiration for everyone.

The publisher and the editorial board hope that this book will prove to be a valuable piece of knowledge for researchers, students, practitioners and scholars across the globe.

List of Contributors

Osama Alkhouli
Caterpillar Inc., Illinois, USA

Victor DeBrunner
University of Oklahoma, School of Electrical and Computer Engineering, Oklahoma, USA

Joseph Havlicek
Florida State University, Electrical and Computer Engineering Department, Florida, USA

Tõnu Trump
Tallinn University of Technology, Estonia

Yang Han
Dept. of Power Electronics, School of Mechatronics Engineering, University of Electronic Science and Technology of China, Chengdu, China

Zhidong Zhao, Yi Luo, Fangqin Ren and Li Zhang
Hangzhou Dianzi University, Hangzhou, China

Changchun Shi
Hangzhou Normal University, Hangzhou, China

Edgar Omar Lopez-Caudana
Tecnológico de Monterrey, Campus Ciudad de México, Mexico

Hector Manuel Perez-Meana
SEPI, ESIME Culhuacan, IPN, Mexico

Printed in the USA
CPSIA information can be obtained
at www.ICGtesting.com
JSHW011809301024
72690JS00002B/8

9 781632 400130